Lecture Notes in Networks and Systems   **988**

The series "Lecture Notes in Networks and Systems" publishes the latest developments in Networks and Systems—quickly, informally and with high quality. Original research reported in proceedings and post-proceedings represents the core of LNNS.

Volumes published in LNNS embrace all aspects and subfields of, as well as new challenges in, Networks and Systems.

The series contains proceedings and edited volumes in systems and networks, spanning the areas of Cyber-Physical Systems, Autonomous Systems, Sensor Networks, Control Systems, Energy Systems, Automotive Systems, Biological Systems, Vehicular Networking and Connected Vehicles, Aerospace Systems, Automation, Manufacturing, Smart Grids, Nonlinear Systems, Power Systems, Robotics, Social Systems, Economic Systems and other. Of particular value to both the contributors and the readership are the short publication timeframe and the worldwide distribution and exposure which enable both a wide and rapid dissemination of research output.

The series covers the theory, applications, and perspectives on the state of the art and future developments relevant to systems and networks, decision making, control, complex processes and related areas, as embedded in the fields of interdisciplinary and applied sciences, engineering, computer science, physics, economics, social, and life sciences, as well as the paradigms and methodologies behind them.

Indexed by SCOPUS, INSPEC, WTI Frankfurt eG, zbMATH, SCImago.

All books published in the series are submitted for consideration in Web of Science.

For proposals from Asia please contact Aninda Bose (aninda.bose@springer.com).

Álvaro Rocha · Hojjat Adeli ·
Gintautas Dzemyda · Fernando Moreira ·
Aneta Poniszewska-Marańda

Editors

Good Practices and New Perspectives in Information Systems and Technologies

WorldCIST 2024, Volume 4

 Springer

Editors
Álvaro Rocha
ISEG
Universidade de Lisboa
Lisbon, Portugal

Hojjat Adeli
College of Engineering
The Ohio State University
Columbus, OH, USA

Gintautas Dzemyda
Institute of Data Science and Digital
Technologies
Vilnius University
Vilnius, Lithuania

Fernando Moreira
DCT
Universidade Portucalense
Porto, Portugal

Aneta Poniszewska-Marańda
Institute of Information Technology
Lodz University of Technology
Łódz, Poland

ISSN 2367-3370   ISSN 2367-3389 (electronic)
Lecture Notes in Networks and Systems
ISBN 978-3-031-60223-8   ISBN 978-3-031-60224-5 (eBook)
https://doi.org/10.1007/978-3-031-60224-5

This Springer imprint is published by the registered company Springer Nature Switzerland AG
The registered company address is: Gewerbestrasse 11, 6330 Cham, Switzerland

If disposing of this product, please recycle the paper.

Preface

This book contains a selection of papers accepted for presentation and discussion at the 2024 World Conference on Information Systems and Technologies (WorldCIST'24). This conference had the scientific support of the Lodz University of Technology, Information and Technology Management Association (ITMA), IEEE Systems, Man, and Cybernetics Society (IEEE SMC), Iberian Association for Information Systems and Technologies (AISTI), and Global Institute for IT Management (GIIM). It took place in Lodz city, Poland, 26–28 March 2024.

The World Conference on Information Systems and Technologies (WorldCIST) is a global forum for researchers and practitioners to present and discuss recent results and innovations, current trends, professional experiences, and challenges of modern Information Systems and Technologies research, technological development, and applications. One of its main aims is to strengthen the drive toward a holistic symbiosis between academy, society, and industry. WorldCIST'24 is built on the successes of: WorldCIST'13 held at Olhão, Algarve, Portugal; WorldCIST'14 held at Funchal, Madeira, Portugal; WorldCIST'15 held at São Miguel, Azores, Portugal; WorldCIST'16 held at Recife, Pernambuco, Brazil; WorldCIST'17 held at Porto Santo, Madeira, Portugal; WorldCIST'18 held at Naples, Italy; WorldCIST'19 held at La Toja, Spain; WorldCIST'20 held at Budva, Montenegro; WorldCIST'21 held at Terceira Island, Portugal; WorldCIST'22 held at Budva, Montenegro; and WorldCIST'23, which took place at Pisa, Italy.

The Program Committee of WorldCIST'24 was composed of a multidisciplinary group of 328 experts and those who are intimately concerned with Information Systems and Technologies. They have had the responsibility for evaluating, in a 'blind review' process, the papers received for each of the main themes proposed for the conference: A) Information and Knowledge Management; B) Organizational Models and Information Systems; C) Software and Systems Modeling; D) Software Systems, Architectures, Applications and Tools; E) Multimedia Systems and Applications; F) Computer Networks, Mobility and Pervasive Systems; G) Intelligent and Decision Support Systems; H) Big Data Analytics and Applications; I) Human-Computer Interaction; J) Ethics, Computers & Security; K) Health Informatics; L) Information Technologies in Education; M) Information Technologies in Radiocommunications; and N) Technologies for Biomedical Applications.

The conference also included workshop sessions taking place in parallel with the conference ones. Workshop sessions covered themes such as: ICT for Auditing & Accounting; Open Learning and Inclusive Education Through Information and Communication Technology; Digital Marketing and Communication, Technologies, and Applications; Advances in Deep Learning Methods and Evolutionary Computing for Health Care; Data Mining and Machine Learning in Smart Cities: The role of the technologies in the research of the migrations; Artificial Intelligence Models and Artifacts for Business Intelligence Applications; AI in Education; Environmental data analytics; Forest-Inspired

Computational Intelligence Methods and Applications; Railway Operations, Modeling and Safety; Technology Management in the Electrical Generation Industry: Capacity Building through Knowledge, Resources and Networks; Data Privacy and Protection in Modern Technologies; Strategies and Challenges in Modern NLP: From Argumentation to Ethical Deployment; and Enabling Software Engineering Practices Via Last Development Trends.

WorldCIST'24 and its workshops received about 400 contributions from 47 countries around the world. The papers accepted for oral presentation and discussion at the conference are published by Springer (this book) in six volumes and will be submitted for indexing by WoS, Scopus, EI-Compendex, DBLP, and/or Google Scholar, among others. Extended versions of selected best papers will be published in special or regular issues of leading and relevant journals, mainly JCR/SCI/SSCI and Scopus/EI-Compendex indexed journals.

We acknowledge all of those that contributed to the staging of WorldCIST'24 (authors, committees, workshop organizers, and sponsors). We deeply appreciate their involvement and support that was crucial for the success of WorldCIST'24.

March 2024

Álvaro Rocha
Hojjat Adeli
Gintautas Dzemyda
Fernando Moreira
Aneta Poniszewska-Marańda

Organization

Conference

Honorary Chair

Hojjat Adeli — The Ohio State University, USA

General Chair

Álvaro Rocha — ISEG, University of Lisbon, Portugal

Co-chairs

Gintautas Dzemyda — Vilnius University, Lithuania
Sandra Costanzo — University of Calabria, Italy

Workshops Chair

Fernando Moreira — Portucalense University, Portugal

Local Organizing Committee

Bożena Borowska — Lodz University of Technology, Poland
Łukasz Chomątek — Lodz University of Technology, Poland
Joanna Ochelska-Mierzejewska — Lodz University of Technology, Poland
Aneta Poniszewska-Marańda — Lodz University of Technology, Poland

Advisory Committee

Ana Maria Correia (Chair) — University of Sheffield, UK
Brandon Randolph-Seng — Texas A&M University, USA

Chris Kimble	KEDGE Business School & MRM, UM2, Montpellier, France
Damian Niwiński	University of Warsaw, Poland
Eugene Spafford	Purdue University, USA
Florin Gheorghe Filip	Romanian Academy, Romania
Janusz Kacprzyk	Polish Academy of Sciences, Poland
João Tavares	University of Porto, Portugal
Jon Hall	The Open University, UK
John MacIntyre	University of Sunderland, UK
Karl Stroetmann	Empirica Communication & Technology Research, Germany
Marjan Mernik	University of Maribor, Slovenia
Miguel-Angel Sicilia	University of Alcalá, Spain
Mirjana Ivanovic	University of Novi Sad, Serbia
Paulo Novais	University of Minho, Portugal
Sami Habib	Kuwait University, Kuwait
Wim Van Grembergen	University of Antwerp, Belgium

Program Committee Co-chairs

Adam Wojciechowski	Lodz University of Technology, Poland
Aneta Poniszewska-Marańda	Lodz University of Technology, Poland

Program Committee

Abderrahmane Ez-zahout	Mohammed V University, Morocco
Adriana Peña Pérez Negrón	Universidad de Guadalajara, Mexico
Adriani Besimi	South East European University, North Macedonia
Agostinho Sousa Pinto	Polytechnic of Porto, Portugal
Ahmed El Oualkadi	Abdelmalek Essaadi University, Morocco
Akex Rabasa	University Miguel Hernandez, Spain
Alanio de Lima	UFC, Brazil
Alba Córdoba-Cabús	University of Malaga, Spain
Alberto Freitas	FMUP, University of Porto, Portugal
Aleksandra Labus	University of Belgrade, Serbia
Alessio De Santo	HE-ARC, Switzerland
Alexandru Vulpe	University Politechnica of Bucharest, Romania
Ali Idri	ENSIAS, University Mohamed V, Morocco
Alicia García-Holgado	University of Salamanca, Spain

Almir Souza Silva Neto	IFMA, Brazil
Álvaro López-Martín	University of Malaga, Spain
Amélia Badica	Universiti of Craiova, Romania
Amélia Cristina Ferreira Silva	Polytechnic of Porto, Portugal
Amit Shelef	Sapir Academic College, Israel
Ana Carla Amaro	Universidade de Aveiro, Portugal
Ana Dinis	Polytechnic of Cávado and Ave, Portugal
Ana Isabel Martins	University of Aveiro, Portugal
Anabela Gomes	University of Coimbra, Portugal
Anacleto Correia	CINAV, Portugal
Andrew Brosnan	University College Cork, Ireland
Andjela Draganic	University of Montenegro, Montenegro
Aneta Polewko-Klim	University of Białystok, Institute of Informatics, Poland
Aneta Poniszewska-Maranda	Lodz University of Technology, Poland
Angeles Quezada	Instituto Tecnologico de Tijuana, Mexico
Anis Tissaoui	University of Jendouba, Tunisia
Ankur Singh Bist	KIET, India
Ann Svensson	University West, Sweden
Anna Gawrońska	Poznański Instytut Technologiczny, Poland
Antoni Oliver	University of the Balearic Islands, Spain
Antonio Jiménez-Martín	Universidad Politécnica de Madrid, Spain
Aroon Abbu	Bell and Howell, USA
Arslan Enikeev	Kazan Federal University, Russia
Beatriz Berrios Aguayo	University of Jaen, Spain
Benedita Malheiro	Polytechnic of Porto, ISEP, Portugal
Bertil Marques	Polytechnic of Porto, ISEP, Portugal
Boris Shishkov	ULSIT/IMI - BAS/IICREST, Bulgaria
Borja Bordel	Universidad Politécnica de Madrid, Spain
Branko Perisic	Faculty of Technical Sciences, Serbia
Bruno F. Gonçalves	Polytechnic of Bragança, Portugal
Carla Pinto	Polytechnic of Porto, ISEP, Portugal
Carlos Balsa	Polytechnic of Bragança, Portugal
Carlos Rompante Cunha	Polytechnic of Bragança, Portugal
Catarina Reis	Polytechnic of Leiria, Portugal
Célio Gonçalo Marques	Polytenic of Tomar, Portugal
Cengiz Acarturk	Middle East Technical University, Turkey
Cesar Collazos	Universidad del Cauca, Colombia
Cristina Gois	Polytechnic University of Coimbra, Portugal
Christophe Guyeux	Universite de Bourgogne Franche Comté, France
Christophe Soares	University Fernando Pessoa, Portugal
Christos Bouras	University of Patras, Greece

Christos Chrysoulas	London South Bank University, UK
Christos Chrysoulas	Edinburgh Napier University, UK
Ciro Martins	University of Aveiro, Portugal
Claudio Sapateiro	Polytechnic of Setúbal, Portugal
Cosmin Striletchi	Technical University of Cluj-Napoca, Romania
Costin Badica	University of Craiova, Romania
Cristian García Bauza	PLADEMA-UNICEN-CONICET, Argentina
Cristina Caridade	Polytechnic of Coimbra, Portugal
Danish Jamil	Malaysia University of Science and Technology, Malaysia
David Cortés-Polo	University of Extremadura, Spain
David Kelly	University College London, UK
Daria Bylieva	Peter the Great St. Petersburg Polytechnic University, Russia
Dayana Spagnuelo	Vrije Universiteit Amsterdam, Netherlands
Dhouha Jaziri	University of Sousse, Tunisia
Dmitry Frolov	HSE University, Russia
Dulce Mourato	ISTEC - Higher Advanced Technologies Institute Lisbon, Portugal
Edita Butrime	Lithuanian University of Health Sciences, Lithuania
Edna Dias Canedo	University of Brasilia, Brazil
Egils Ginters	Riga Technical University, Latvia
Ekaterina Isaeva	Perm State University, Russia
Eliana Leite	University of Minho, Portugal
Enrique Pelaez	ESPOL University, Ecuador
Eriks Sneiders	Stockholm University, Sweden; Esteban Castellanos ESPE, Ecuador
Fatima Azzahra Amazal	Ibn Zohr University, Morocco
Fernando Bobillo	University of Zaragoza, Spain
Fernando Molina-Granja	National University of Chimborazo, Ecuador
Fernando Moreira	Portucalense University, Portugal
Fernando Ribeiro	Polytechnic Castelo Branco, Portugal
Filipe Caldeira	Polytechnic of Viseu, Portugal
Filippo Neri	University of Naples, Italy
Firat Bestepe	Republic of Turkey Ministry of Development, Turkey
Francesco Bianconi	Università degli Studi di Perugia, Italy
Francisco García-Peñalvo	University of Salamanca, Spain
Francisco Valverde	Universidad Central del Ecuador, Ecuador
Frederico Branco	University of Trás-os-Montes e Alto Douro, Portugal
Galim Vakhitov	Kazan Federal University, Russia

Gayo Diallo	University of Bordeaux, France
Gabriel Pestana	Polytechnic Institute of Setubal, Portugal
Gema Bello-Orgaz	Universidad Politecnica de Madrid, Spain
George Suciu	BEIA Consult International, Romania
Ghani Albaali	Princess Sumaya University for Technology, Jordan
Gian Piero Zarri	University Paris-Sorbonne, France
Giovanni Buonanno	University of Calabria, Italy
Gonçalo Paiva Dias	University of Aveiro, Portugal
Goreti Marreiros	ISEP/GECAD, Portugal
Habiba Drias	University of Science and Technology Houari Boumediene, Algeria
Hafed Zarzour	University of Souk Ahras, Algeria
Haji Gul	City University of Science and Information Technology, Pakistan
Hakima Benali Mellah	Cerist, Algeria
Hamid Alasadi	Basra University, Iraq
Hatem Ben Sta	University of Tunis at El Manar, Tunisia
Hector Fernando Gomez Alvarado	Universidad Tecnica de Ambato, Ecuador
Hector Menendez	King's College London, UK
Hélder Gomes	University of Aveiro, Portugal
Helia Guerra	University of the Azores, Portugal
Henrique da Mota Silveira	University of Campinas (UNICAMP), Brazil
Henrique S. Mamede	University Aberta, Portugal
Henrique Vicente	University of Évora, Portugal
Hicham Gueddah	University Mohammed V in Rabat, Morocco
Hing Kai Chan	University of Nottingham Ningbo China, China
Igor Aguilar Alonso	Universidad Nacional Tecnológica de Lima Sur, Peru
Inês Domingues	University of Coimbra, Portugal
Isabel Lopes	Polytechnic of Bragança, Portugal
Isabel Pedrosa	Coimbra Business School - ISCAC, Portugal
Isaías Martins	University of Leon, Spain
Issam Moghrabi	Gulf University for Science and Technology, Kuwait
Ivan Armuelles Voinov	University of Panama, Panama
Ivan Dunđer	University of Zagreb, Croatia
Ivone Amorim	University of Porto, Portugal
Jaime Diaz	University of La Frontera, Chile
Jan Egger	IKIM, Germany
Jan Kubicek	Technical University of Ostrava, Czech Republic
Jeimi Cano	Universidad de los Andes, Colombia

Jesús Gallardo Casero	University of Zaragoza, Spain
Jezreel Mejia	CIMAT, Unidad Zacatecas, Mexico
Jikai Li	The College of New Jersey, USA
Jinzhi Lu	KTH-Royal Institute of Technology, Sweden
Joao Carlos Silva	IPCA, Portugal
João Manuel R. S. Tavares	University of Porto, FEUP, Portugal
João Paulo Pereira	Polytechnic of Bragança, Portugal
João Reis	University of Aveiro, Portugal
João Reis	University of Lisbon, Portugal
João Rodrigues	University of the Algarve, Portugal
João Vidal de Carvalho	Polytechnic of Porto, Portugal
Joaquin Nicolas Ros	University of Murcia, Spain
John W. Castro	University de Atacama, Chile
Jorge Barbosa	Polytechnic of Coimbra, Portugal
Jorge Buele	Technical University of Ambato, Ecuador; Jorge Gomes University of Lisbon, Portugal
Jorge Oliveira e Sá	University of Minho, Portugal
José Braga de Vasconcelos	Universidade Lusófona, Portugal
Jose M. Parente de Oliveira	Aeronautics Institute of Technology, Brazil
José Machado	University of Minho, Portugal
José Paulo Lousado	Polytechnic of Viseu, Portugal
Jose Quiroga	University of Oviedo, Spain
Jose Silvestre Silva	Academia Military, Portugal
Jose Torres	University Fernando Pessoa, Portugal
Juan M. Santos	University of Vigo, Spain
Juan Manuel Carrillo de Gea	University of Murcia, Spain
Juan Pablo Damato	UNCPBA-CONICET, Argentina
Kalinka Kaloyanova	Sofia University, Bulgaria
Kamran Shaukat	The University of Newcastle, Australia
Katerina Zdravkova	University Ss. Cyril and Methodius, North Macedonia
Khawla Tadist	Morocco
Khalid Benali	LORIA - University of Lorraine, France
Khalid Nafil	Mohammed V University in Rabat, Morocco
Korhan Gunel	Adnan Menderes University, Turkey
Krzysztof Wolk	Polish-Japanese Academy of Information Technology, Poland
Kuan Yew Wong	Universiti Teknologi Malaysia (UTM), Malaysia
Kwanghoon Kim	Kyonggi University, South Korea
Laila Cheikhi	Mohammed V University in Rabat, Morocco
Laura Varela-Candamio	Universidade da Coruña, Spain
Laurentiu Boicescu	E.T.T.I. U.P.B., Romania

Lbtissam Abnane	ENSIAS, Morocco
Lia-Anca Hangan	Technical University of Cluj-Napoca, Romania
Ligia Martinez	CECAR, Colombia
Lila Rao-Graham	University of the West Indies, Jamaica
Liliana Ivone Pereira	Polytechnic of Cávado and Ave, Portugal
Łukasz Tomczyk	Pedagogical University of Cracow, Poland
Luis Alvarez Sabucedo	University of Vigo, Spain
Luís Filipe Barbosa	University of Trás-os-Montes e Alto Douro
Luis Mendes Gomes	University of the Azores, Portugal
Luis Pinto Ferreira	Polytechnic of Porto, Portugal
Luis Roseiro	Polytechnic of Coimbra, Portugal
Luis Silva Rodrigues	Polytencic of Porto, Portugal
Mahdieh Zakizadeh	MOP, Iran
Maksim Goman	JKU, Austria
Manal el Bajta	ENSIAS, Morocco
Manuel Antonio Fernández-Villacañas Marín	Technical University of Madrid, Spain
Manuel Ignacio Ayala Chauvin	University Indoamerica, Ecuador
Manuel Silva	Polytechnic of Porto and INESC TEC, Portugal
Manuel Tupia	Pontifical Catholic University of Peru, Peru
Manuel Au-Yong-Oliveira	University of Aveiro, Portugal
Marcelo Mendonça Teixeira	Universidade de Pernambuco, Brazil
Marciele Bernardes	University of Minho, Brazil
Marco Ronchetti	Universita' di Trento, Italy
Mareca María Pilar	Universidad Politécnica de Madrid, Spain
Marek Kvet	Zilinska Univerzita v Ziline, Slovakia
Maria João Ferreira	Universidade Portucalense, Portugal
Maria José Sousa	University of Coimbra, Portugal
María Teresa García-Álvarez	University of A Coruna, Spain
Maria Sokhn	University of Applied Sciences of Western Switzerland, Switzerland
Marijana Despotovic-Zrakic	Faculty Organizational Science, Serbia
Marilio Cardoso	Polytechnic of Porto, Portugal
Mário Antunes	Polytechnic of Leiria & CRACS INESC TEC, Portugal
Marisa Maximiano	Polytechnic Institute of Leiria, Portugal
Marisol Garcia-Valls	Polytechnic University of Valencia, Spain
Maristela Holanda	University of Brasilia, Brazil
Marius Vochin	E.T.T.I. U.P.B., Romania
Martin Henkel	Stockholm University, Sweden
Martín López Nores	University of Vigo, Spain
Martin Zelm	INTEROP-VLab, Belgium

Mazyar Zand	MOP, Iran
Mawloud Mosbah	University 20 Août 1955 of Skikda, Algeria
Michal Adamczak	Poznan School of Logistics, Poland
Michal Kvet	University of Zilina, Slovakia
Miguel Garcia	University of Oviedo, Spain
Mircea Georgescu	Al. I. Cuza University of Iasi, Romania
Mirna Muñoz	Centro de Investigación en Matemáticas A.C., Mexico
Mohamed Hosni	ENSIAS, Morocco
Monica Leba	University of Petrosani, Romania
Nadesda Abbas	UBO, Chile
Narasimha Rao Vajjhala	University of New York Tirana, Tirana
Narjes Benameur	Laboratory of Biophysics and Medical Technologies of Tunis, Tunisia
Natalia Grafeeva	Saint Petersburg University, Russia
Natalia Miloslavskaya	National Research Nuclear University MEPhI, Russia
Naveed Ahmed	University of Sharjah, United Arab Emirates
Neeraj Gupta	KIET group of institutions Ghaziabad, India
Nelson Rocha	University of Aveiro, Portugal
Nikola S. Nikolov	University of Limerick, Ireland
Nicolas de Araujo Moreira	Federal University of Ceara, Brazil
Nikolai Prokopyev	Kazan Federal University, Russia
Niranjan S. K.	JSS Science and Technology University, India
Noemi Emanuela Cazzaniga	Politecnico di Milano, Italy
Noureddine Kerzazi	Polytechnique Montréal, Canada
Nuno Melão	Polytechnic of Viseu, Portugal
Nuno Octávio Fernandes	Polytechnic of Castelo Branco, Portugal
Nuno Pombo	University of Beira Interior, Portugal
Olga Kurasova	Vilnius University, Lithuania
Olimpiu Stoicuta	University of Petrosani, Romania
Patricia Quesado	Polytechnic of Cávado and Ave, Portugal
Patricia Zachman	Universidad Nacional del Chaco Austral, Argentina
Paula Serdeira Azevedo	University of Algarve, Portugal
Paula Dias	Polytechnic of Guarda, Portugal
Paulo Alejandro Quezada Sarmiento	University of the Basque Country, Spain
Paulo Maio	Polytechnic of Porto, ISEP, Portugal
Paulvanna Nayaki Marimuthu	Kuwait University, Kuwait
Paweł Karczmarek	The John Paul II Catholic University of Lublin, Poland

Pedro Rangel Henriques	University of Minho, Portugal
Pedro Sobral	University Fernando Pessoa, Portugal
Pedro Sousa	University of Minho, Portugal
Philipp Jordan	University of Hawaii at Manoa, USA
Piotr Kulczycki	Systems Research Institute, Polish Academy of Sciences, Poland
Prabhat Mahanti	University of New Brunswick, Canada
Rabia Azzi	Bordeaux University, France
Radu-Emil Precup	Politehnica University of Timisoara, Romania
Rafael Caldeirinha	Polytechnic of Leiria, Portugal
Raghuraman Rangarajan	Sequoia AT, Portugal
Radhakrishna Bhat	Manipal Institute of Technology, India
Raiani Ali	Hamad Bin Khalifa University, Qatar
Ramadan Elaiess	University of Benghazi, Libya
Ramayah T.	Universiti Sains Malaysia, Malaysia
Ramazy Mahmoudi	University of Monastir, Tunisia
Ramiro Gonçalves	University of Trás-os-Montes e Alto Douro & INESC TEC, Portugal
Ramon Alcarria	Universidad Politécnica de Madrid, Spain
Ramon Fabregat Gesa	University of Girona, Spain
Ramy Rahimi	Chungnam National University, South Korea
Reiko Hishiyama	Waseda University, Japan
Renata Maria Maracho	Federal University of Minas Gerais, Brazil
Renato Toasa	Israel Technological University, Ecuador
Reyes Juárez Ramírez	Universidad Autonoma de Baja California, Mexico
Rocío González-Sánchez	Rey Juan Carlos University, Spain
Rodrigo Franklin Frogeri	University Center of Minas Gerais South, Brazil
Ruben Pereira	ISCTE, Portugal
Rui Alexandre Castanho	WSB University, Poland
Rui S. Moreira	UFP & INESC TEC & LIACC, Portugal
Rustam Burnashev	Kazan Federal University, Russia
Saeed Salah	Al-Quds University, Palestine
Said Achchab	Mohammed V University in Rabat, Morocco
Sajid Anwar	Institute of Management Sciences Peshawar, Pakistan
Sami Habib	Kuwait University, Kuwait
Samuel Sepulveda	University of La Frontera, Chile
Sara Luis Dias	Polytechnic of Cávado and Ave, Portugal
Sandra Costanzo	University of Calabria, Italy
Sandra Patricia Cano Mazuera	University of San Buenaventura Cali, Colombia
Sassi Sassi	FSJEGJ, Tunisia

Seppo Sirkemaa	University of Turku, Finland
Sergio Correia	Polytechnic of Portalegre, Portugal
Shahnawaz Talpur	Mehran University of Engineering & Technology Jamshoro, Pakistan
Shakti Kundu	Manipal University Jaipur, Rajasthan, India
Shashi Kant Gupta	Eudoxia Research University, USA
Silviu Vert	Politehnica University of Timisoara, Romania
Simona Mirela Riurean	University of Petrosani, Romania
Slawomir Zolkiewski	Silesian University of Technology, Poland
Solange Rito Lima	University of Minho, Portugal
Sonia Morgado	ISCPSI, Portugal
Sonia Sobral	Portucalense University, Portugal
Sorin Zoican	Polytechnic University of Bucharest, Romania
Souraya Hamida	Batna 2 University, Algeria
Stalin Figueroa	University of Alcala, Spain
Sümeyya Ilkin	Kocaeli University, Turkey
Syed Asim Ali	University of Karachi, Pakistan
Syed Nasirin	Universiti Malaysia Sabah, Malaysia
Tatiana Antipova	Institute of Certified Specialists, Russia
Tatianna Rosal	University of Trás-os-Montes e Alto Douro, Portugal
Tero Kokkonen	JAMK University of Applied Sciences, Finland
The Thanh Van	HCMC University of Food Industry, Vietnam
Thomas Weber	EPFL, Switzerland
Timothy Asiedu	TIM Technology Services Ltd., Ghana
Tom Sander	New College of Humanities, Germany
Tomasz Kisielewicz	Warsaw University of Technology
Tomaž Klobučar	Jozef Stefan Institute, Slovenia
Toshihiko Kato	University of Electro-communications, Japan
Tuomo Sipola	Jamk University of Applied Sciences, Finland
Tzung-Pei Hong	National University of Kaohsiung, Taiwan
Valentim Realinho	Polytechnic of Portalegre, Portugal
Valentina Colla	Scuola Superiore Sant'Anna, Italy
Valerio Stallone	ZHAW, Switzerland
Verónica Vasconcelos	Polytechnic of Coimbra, Portugal
Vicenzo Iannino	Scuola Superiore Sant'Anna, Italy
Vitor Gonçalves	Polytechnic of Bragança, Portugal
Victor Alves	University of Minho, Portugal
Victor Georgiev	Kazan Federal University, Russia
Victor Hugo Medina Garcia	Universidad Distrital Francisco José de Caldas, Colombia
Victor Kaptelinin	Umeå University, Sweden

Viktor Medvedev	Vilnius University, Lithuania
Vincenza Carchiolo	University of Catania, Italy
Waqas Bangyal	University of Gujrat, Pakistan
Wolf Zimmermann	Martin Luther University Halle-Wittenberg, Germany
Yadira Quiñonez	Autonomous University of Sinaloa, Mexico
Yair Wiseman	Bar-Ilan University, Israel
Yassine Drias	University of Algiers, Algeria
Yuhua Li	Cardiff University, UK
Yuwei Lin	University of Roehampton, UK
Zbigniew Suraj	University of Rzeszow, Poland
Zorica Bogdanovic	University of Belgrade, Serbia

Contents

Information Technologies in Radiocommunications

Information Technologies in Education

Heuristics for Designing Pervasive Game Experiences in the Older Adult Population

Johnny Salazar Cardona[1], Francisco Luis Gutiérrez Vela[1], Jeferson Arango Lopez[2], and Fernando Moreira[3,4(✉)]

[1] Departamento de Lenguajes y Sistemas Informáticos, ETSI Informática, Universidad de Granada, 18071 Granada, Spain
jasalazar@correo.ugr.es, fgutierr@ugr.es

[2] Departamento de Sistemas e Informática, Facultad de Ingenierías, Universidad de Caldas, Calle 65 # 26-10, Edificio del Parque, Manizales, Caldas, Colombia
jeferson.arango@ucaldas.edu.co

[3] REMIT, IJP, Universidade Portucalense, Rua Dr. António Bernardino Almeida, 541-619, 4200-072 Porto, Portugal
fmoreira@upt.pt

[4] IEETA, Universidade de Aveiro, Aveiro, Portugal

Abstract. Older adults face significant challenges when trying to incorporate contemporary technology into their lives, which negatively impacts their ability to take full advantage of technological advances in their daily routines. This problem is particularly evident in the entertainment domain, where technological video games can be beneficial to those who venture to use them. Although a growing number of older people are adopting digital games in their daily activities, increasing technological sophistication also creates obstacles in their adaptation process. An example of this is the proliferation of pervasive technological games, which make use of devices such as virtual assistants, sensors embedded in mobile devices, and virtual reality to provide advantages to the older adult population. This article will address the fundamental aspects aimed at improving the game experience in these pervasive environments, taking into consideration the specific needs of this demographic group. This results in the specification of a set of heuristics that can be used to identify potential design problems in game-based systems with some degree of pervasiveness oriented to the elderly population.

Keywords: Older Adults · Pervasive · Game Based System · Player Experience · Playability

1 Introduction

The unfamiliarity that older adults experience with technology prevents them from taking full advantage of the benefits it can offer. This limitation is evident in different situations, such as leisure and fun time. In this sense, it has been shown that technological games have a positive impact on the older adult population that has become familiar with them. Among the games that are particularly effective for this group are those that involve direct and natural interaction, stimulating multiple senses at simultaneously. Pervasive games, an emerging genre, have great potential to generate positive experiences and emotions in older adults [1]. These games can not only be entertaining, but also offer several benefits. They can be easily integrated into the daily routine, encouraging physical activity, cognitive training and socialization, which contributes to the overall well-being of this population group.

Some pervasive games incorporate physical locations, stories and social contexts into their game dynamics, which can be beneficial for older adults in terms of improving their quality of life, especially in physical aspects such as walking and specific exercises [2]. These games are characterized by being more relaxed, avoiding cognitive overload on players and facilitating their learning and adaptation process. In addition, these games have the advantages of fostering social connections, increasing physical activity and improving mental well-being [3]. They also focus on providing flexible experiences for casual gamers, allowing seniors to learn at their own pace and participate according to their preferences.

In this study, an approach to design immersive experiences for older adults is presented. This model has been formulated using specific heuristics and checklists, with the purpose of simplifying the creation of game systems based on the notion of "pervasiveness" and providing an enriching game experience, taking into consideration the needs of this demographic group. The text is structured as follows: In Sect. 2, the concept of pervasivity is contextualized and examples of its previous application in the older adult population are provided. It also reviews our base proposal of transversal elements that are essential to guarantee an optimal game experience by facilitating the interaction between the pervasive experience and older adults; In Sect. 3, the methodological process employed to define the transversal elements model and its conversion into applicable heuristics is detailed; Finally, in Sect. 4, an analysis of the results obtained is carried out, the conclusions of the study are presented, and possible directions for future research are raised.

2 Background

The concept of "pervasiveness" has found application in the field of entertainment, particularly in the context of pervasive games. Games of this kind are characterized by being playful experiences that merge with the tangible environment, thus eliminating the boundaries that have historically defined the game [4]. This means that the game is no longer limited to a specific place, time or number of players. Pervasive games transcend traditional space, social, contextual and temporal boundaries, expanding the game space and offering a richer, more immersive game experience [5].

Pervasive games are distinguished by their ability to combine game rules with elements of the real environment, such as geographical locations, individuals and tangible objects. This fusion generates a singular game experience that transcends the conventional barriers that usually separate the player from reality, generating a greater immersion in the game experience. Additionally, these games promote a robust social dimension, characterized by natural interactions ranging from voice communication to gestures, actions and the manipulation of physical objects. One of the primary benefits of pervasive games is their aptitude to address the challenges inherent in conventional digital games. These games promote both physical activity and social interaction, while linking the intrinsic motivation of the participants [6].

In the older adult population, a variety of pervasive game experiences have been developed. Some notable examples include titles such as "Age Invaders" and "The Fantastic Journey" [7]. In addition to games aimed at the general public, there are also specific initiatives for older adults. These include escape games such as "A Tale of Tales" and location-based collection games in open spaces, such as "Life Chasing" and "Shinpo" [1]. These games are mainly used to promote physical activity, cognitive training and social interaction in this demographic. Several exergaming-type game applications have been developed that target the older population. These applications include "Dancetown Fitness System", "Smart Chair", "Wii bowling", "Table tennis game", "Basketball Genius", "Flying Eagle", "Ping Pong", "Escape Room", "Safari Move" and "SportWall".

To generate pervasive game experiences that meet the needs of the elderly population, it is necessary to consider the physical and cognitive characteristics of natural human aging. To achieve this goal, a previous systematic review [8] was conducted to analyze the use of pervasive game experiences in elderly individuals. This process yielded detailed information about the various experiences provided through systems based on pervasive games for this demographic segment.

2.1 Proposed Model: Pervasiveness Pyramid

After conducting the exhaustive systematic review [8], identified a series of elements of primary consideration in the design of pervasive game experiences oriented towards the older adult population. These findings were used to define a model called the " pervasiveness pyramid", which emerged as an effective means to evaluate the degree of pervasiveness provided by a game experience [9] (see Fig. 1). In addition, we identified a set of transversal elements that deserve special attention in the context of the older adult population. It is important to note that these elements can be easily adapted and extended to any type of pervasive game-based system and target population [10].

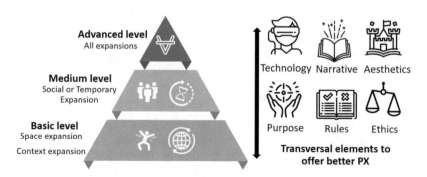

Fig. 1. Pervasiveness pyramid and its transversal elements [9]

The "pervasiveness pyramid" presents a conceptual structure comprising three levels of integration in the context of pervasive game experiences. The initial level addresses the integration of physical and virtual environments using technologies in a specific context, which influences the game experience. The intermediate level focuses on the incorporation of space and social dimensions into the game environment. Finally, the advanced level encompasses the combination of all the above elements. However, up to this point, it has been outlined what aspects should be included in a pervasive experience in a general way, but no guidance has been provided on how to carry out such integration. The methodology to achieve this is based on the adaptation of the transversal elements identified, which must be adjusted according to the nature of the pervasive game system and its target audience, in this case, the older adult population. These identified elements focus on technology-based pervasive game systems, with the primary objective of offering an enriching game experience. The elements identified encompass aspects such as the technology employed, the game narrative, the aesthetics, the purpose of the game, the rules governing the game and, crucially, the ethical components guiding its development and application.

Ultimately, in the conceptualization of this model, it became necessary to specify different properties for the various expansions of pervasivity. These properties played a key role in the subsequent formulation of heuristics and checklists intended to provide a means by which they can be applied in the detection of potential design problems and in the evaluation of playability.

3 Materials and Methods

To provide an exhaustive description of the generation of the explained "pervasiveness pyramid" model and the transversal elements that should be considered to optimize the game experience in game-based systems (GBS) with some degree of pervasiveness aimed at the elderly population, a rigorous search and definition process was carried out using a methodology designed to put the proposed theoretical model into practice. To achieve this purpose, the heuristic definition methodology proposed by Quiñonez et al. [11], which details a formal process of heuristic definition, was applied. This methodological proposal consists of a series of stages that guide the process, each with its respective inputs and outputs. These stages are divided into the following categories: exploratory, experimental, descriptive, correlational, selection and specification. As a result of the implementation of this methodology, a set of heuristics was generated.

3.1 Exploratory and Experimental Stage

An exploratory process has been carried out through a systematic review with the main purpose of acquiring knowledge about the enjoyment of the elderly in game experiences, especially those of a pervasive nature [8]. In addition, information has been compiled on the current strategies employed for the implementation of these game experiences, the degree of acceptance shown by older adults towards them, as well as the different mechanics and dynamics applied specifically for this demographic group.

To the systematic review, the methodology established by Kitchenham and Charters was used [12], which establishes a set of steps or stages that guide the implementation of systematic reviews in the field of software. In terms of article selection, a search string was designed using logical operators and relevant terms to effectively filter the desired results. In addition, inclusion and exclusion criteria were specified to reduce the total number of articles to be considered in the elaboration of the proposal presented here. It should be noted that the data corresponding to the experimental phase were obtained through a rigorous systematic review process, eliminating the need to carry out additional experiments to obtain the required information.

3.2 Descriptive Stage

The findings derived from the systematic review process indicate the following: First, the important application in the stimulation of physical and cognitive activities, as well as in the generation of feelings of wellbeing at a social level and in the field of learning [8]. Second, distinctive features and particular characteristics inherent to these game experiences have been identified. These elements have served as a foundation in the formulation of essential considerations to be considered when designing pervasive experiences aimed at this population. Third, it is worth noting that the results obtained by Rienzo and Cubillos in a systematic review of a similar scope [13], although not focused on pervasive experiences specifically, but on game experiences, have also been considered. In both reviews, heuristics related to playability and player experience (PX) have been identified, as well as heuristics concerning usability and user experience (UX) in GBS.

Fourth, about previous research on pervasive game experiences specifically aimed at older adults, it is evident that there are no models, concrete guidelines or heuristic principles for the evaluation of playability in this context. However, there are instances of implementation of pervasive game experiences with commercial or therapeutic objectives [7].

Fifth, in designing appropriate game experiences for the older adult population, the incorporation of diverse dynamics, including physical activities, cognitive training exercises, and the implementation of immersive and non-immersive experiences stand out. These practices have been addressed in notable previous research, including studies conducted by I. Awada [14] and G. de Paula [15]. Sixth, regarding the mechanics used in these experiences, there is a high degree of implementation of elements such as challenges, the experiences offered, the accumulation of points to evaluate the performance of participants, as well as the introduction of specific themes to enrich the experience of older adults. Some examples of research that have addressed these mechanics include the work of E. Seah [16] and M. Tabak [17].

3.3 Correlational and Selection Stage

Several elements have been identified that should be considered in the design of pervasive game experiences for older adults. These elements include aspects related to the aesthetics of the game experience, the narrative, the purpose, the rules, the ethics, and the technology that will be used to facilitate the pervasive experience.

To verify that these elements cover the various characteristics of pervasiveness, correlations were carried out with each of the heuristic proposals and selected findings. These correlations have revealed that the proposed elements have sufficient scope to address the reference characteristics, in addition to addressing aspects that have not been considered in the existing definitions.

No specific sets of heuristics oriented toward the design of pervasive game experiences targeting older adults were identified. However, playability heuristics were identified that have been applied in the context of game experiences on mobile devices targeting this demographic. In addition, playability heuristics targeted at a general audience, that is, designed to meet the needs of a diverse set of gamers, have been identified, although these have not been specifically applied in the context of pervasive experiences. Likewise, heuristics related to user experience and usability have been documented, although these have been mainly oriented towards transactional systems.

Due to the absence of specific heuristic guidelines, the decision was made to formulate a set of heuristics of our own. These were elaborated considering the previously exposed findings and the elements identified as key in the design of pervasive experiences aimed at the older adult population (see Table 1).

Table 1. Key transversal elements definition

Proposed key element	Pervasive property covered	Justification
Aesthetics	Virtualization, awareness	The experience of the older adult population is optimized when an environment is created that is aesthetically appealing and facilitates interaction
Narrative	Evolution, Integration	The expanded narrative in pervasivity generates dynamism in the game experience
Technology	Mobility, tangibility, persistence, evolution, surveillance, transmediality	Technological devices for older adults should be designed with their specific needs and capabilities in consideration
Purpose	Social experience, social interaction, mediation, participation	The pervasive experience can serve a variety of purposes, such as entertainment, health improvement, promotion of learning or socialization
Rules	Integration, pop-up gameplay	The dynamics and mechanics of the game must be adjusted to the particularities of the older adult population to offer a safe and adequate game space

(continued)

Table 1. (*continued*)

Proposed key element	Pervasive property covered	Justification
Ethics	Mobility, mediation	Games for older adults should ensure their physical and cognitive health, promote respect and culture, avoid negative emotional states and not be used for hidden purposes

3.4 Specification Stage

Based on the 6 transversal elements identified, the different heuristics that could be used as an evaluation tool were structured. For each heuristic, an identification code, a name, a priority and a basic definition were assigned. Their purpose, the specific characteristic of the pervasive experience they affect, an example of application and the benefits for the elderly population were explained. Finally, we addressed possible interpretation problems, proposed a guiding checklist, indicated to which transversal element of the defined model it corresponds, associated the pervasive property addressed, and indicated to which playability attribute and playability facet it corresponds (see Appendix A). A summarized version of the heuristic specification list can be seen in the Table 2.

Table 2. Summary of defined heuristics

ID	Heuristics	Definition
PH01	Aesthetically pleasing and minimalist visual environments	The pervasive game experience must offer simple, clear and attractive environments to facilitate interaction with the game and generate positive feelings
PH02	Immersive sounds that represent the actions performed	A pervasive game experience that is effective for older adults should use immersive sounds that generate familiar memories and emotions. These sounds can help older adults feel connected to the game and understand their progress
PH03	Secure and adjustable interaction	Immersive game experiences must be safe, appropriate and specific to the needs and characteristics of older adults
PH04	Continued support and feedback to guide and instruct the older adult	The pervasive game experience should provide the older adult with clear, consistent guidance and opportunities for feedback. This will help ensure that older adults can enjoy the game experience without obstacles

(*continued*)

Table 2. (*continued*)

ID	Heuristics	Definition
PH05	Provide purpose and highlight the benefits of the game experience	The pervasive game experience should offer the older adult an environment that generates well-being and positive feelings, as well as fun. The benefits of using this experience should be highlighted clearly and concisely
PH06	Simple rules and balanced difficulty	Experiences should be offered that require short periods of time, with simple regulation, with balanced difficulty, and with the possibility of individual or social interaction
PH07	Pleasant and easy-to-use pervasive technologies	To offer pervasive experiences, technological devices and peripherals must be direct and natural input, with few buttons, portable, effortless, immersive and easy to configure
PH08	Extended narrative with meaning	The extended narrative should provide a clear and coherent context for the player's actions. It should also generate positive emotions, such as surprise, expectation and empathy. In addition, it should encourage curiosity and learning, and provide the player with a sense of recognition and usefulness for their actions
PH09	Ethics and safety in the game experience	The pervasive game experience must provide the older adult with a safe and respectful environment, both physically and cognitively. This environment must protect their data and states of mind, avoiding any kind of exploitation or abuse

Based on the set of heuristics specified, the PL/PX web platform [18] has been defined for the design and evaluation of game experiences aimed at the older adult population, where the pervasive component is very important. Furthermore, the design of a pervasive game experience with tangible interaction has been carried out using the proposed heuristics [19], and the design has been evaluated through the PL/PX platform.

4 Conclusions and Future Work

Technological GBS with some degree of pervasiveness offer numerous opportunities for the older adult population, but their design and implementation must consider their particularities. To achieve this, it is essential to fully understand the possibilities offered by pervasivity in terms of key dimensions, features and elements such as narrative, aesthetics, technology used, game rules, purpose and ethics. These elements make it possible to offer more immersive and meaningful game experiences for participants, moving the game from the virtual world to the real world, with interesting themes, a deep narrative and a fluid interaction with the game.

The use of a standardized methodology for the objective definition of the heuristics has allowed a clear definition of the proposal proposed. Although the methodological process of reference also involves the validation and refinement stages, this has not yet been carried out, but the progress achieved will be very useful as initial design guidelines for future developments.

As future work, we expect to continue to implement the validation and refinement stage of the reference methodological process to provide a robust means by which GBS can be designed with the unique characteristics of the older adult population in mind. This will allow for a more seamless interaction with GBS and significantly improve the player experience. It is also hoped to continue with a meaningful integration between the characteristics of the game experience and the motivations of older adults, using playability assessment instruments specific to this population, which are simple, intuitive and enjoyable, with the objective of evaluating the game as a product and the fun experienced by the participants.

Acknowledgments. This work has been supported by the PLEISAR-Social project, Ref. PID2022-136779OB-C33, funded by the Ministry of Science and Innovation (MCI/AEI/FEDER, UE). Thanks are due to PhD Alexandru Cristian Rusu for guiding and directly corroborating the proper application of his methodology for the definition of our set of heuristics. Finally, we would like to thank MINCIENCIAS of government of Colombia for providing the necessary resources for the completion of this research.

Appendix A

Heuristics specifications is available in: https://1drv.ms/b/s!AkaBCG8z281S40QZWf PCCQ6gnAYy?e=rjzKTe.

References

1. Santos, L.H., et al.: Pervasive game design to evaluate social interaction effects on levels of physical activity among older adults. J. Rehabil. Assistive Technol. Eng. **6**, 205566831984444 (2019). https://doi.org/10.1177/2055668319844443
2. Vagetti, G.C., Barbosa Filho, V.C., Moreira, N.B., de Oliveira, V., Mazzardo, O., de Campos, W.: Association between physical activity and quality of life in the elderly: a systematic review, 2000–2012. Rev. Bras. Psiquiatr. **36**, 76–88 (2014). https://doi.org/10.1590/1516-4446-2012-0895
3. Courtin, E., Knapp, M.: Social isolation, loneliness and health in old age: a scoping review. Health Soc. Care Community **25**, 799–812 (2017). https://doi.org/10.1111/hsc.12311
4. Arango-López, J., Gutiérrez Vela, F.L., Collazos, C.A., Gallardo, J., Moreira, F.: GeoPGD: methodology for the design and development of geolocated pervasive games. Univ. Access Inf. Soc. **20**, 465–477 (2020). https://doi.org/10.1007/s10209-020-00769-w

5. Huizinga, J.: Homo Ludens: a study of the play-element in culture. Eur. Early Childhood Educ. Res. J. **19**, 1–24 (1938). https://doi.org/10.1177/0907568202009004005
6. Benford, S., Magerkurth, C., Ljungstrand, P.: Bridging the physical and digital in pervasive gaming. Commun. ACM **48**, 54–57 (2005). https://doi.org/10.1145/1047671.1047704
7. Cerezo, E., Blasco, A.C.: The space journey game: an intergenerational pervasive experience. In: Conference on Human Factors in Computing Systems – Proceedings, pp. 1–6 (2019). https://doi.org/10.1145/3290607.3313055
8. Salazar, J.A., Arango, J., Gutiérrez, F.L., Moreira, F.: Older adults and games from a perspective of playability, game experience and pervasive environments: a systematics literature review. World Conf. Inf. Syst. Technol. **2022**, 444–453 (2022). https://doi.org/10.1007/978-3-031-04819-7_42
9. Salazar, J., López, J., Gutiérrez, F., Trillo, J.: Design of technology-based pervasive gaming experiences : properties and degrees of pervasiveness. Congreso Español de Videojuegos 2022 (CEV). (2022)
10. Salazar, J., Arango, J., Gutierrez, F.: Pervasiveness for learning in serious games applied to older adults. In: Technological Ecosystems for Enhancing Multiculturality – TEEM, vol. 1 (2022)
11. Quiñones, D., Rusu, C., Rusu, V.: A methodology to develop usability/user experience heuristics. Comput. Stand. Interfaces **59**, 109–129 (2018). https://doi.org/10.1016/j.csi.2018.03.002
12. Kitchenham, B., Charters, S.: Guidelines for performing systematic literature reviews in software engineering. Engineering **2**, 1051 (2007). https://doi.org/10.1145/1134285.1134500
13. Rienzo, A., Cubillos, C.: Playability and player experience in digital games for elderly: a systematic literature review. Sensors (Switzerland) **20**, 1–23 (2020). https://doi.org/10.3390/s20143958
14. Awada, I.A., Mocanu, I., Florea, A.M., Rusu, L., Arba, R., Cramariuc, B.: Enhancing the physical activity of older adults based on user profiles. In: 16th Networking in Education and Research RoEduNet International Conference, RoEduNet 2017 - Proceedings (2017). https://doi.org/10.1109/ROEDUNET.2017.8123749
15. De Paula, G., Valentim, P., Seixas, F., Santana, R., Muchaluat-Saade, D.: Sensory effects in cognitive exercises for elderly users: Stroop game. In: Proceedings - IEEE Symposium on Computer-Based Medical Systems. 2020-July, 132–137 (2020). https://doi.org/10.1109/CBMS49503.2020.00032
16. Seah, E.T.W., Kaufman, D., Sauvé, L., Zhang, F.: Play, learn, connect: older adults' experience with a multiplayer, educational, digital bingo game. J. Educ. Comput. Res. **56**, 675–700 (2018). https://doi.org/10.1177/0735633117722329
17. Tabak, M., De Vette, F., Van DIjk, H., Vollenbroek-Hutten, M.: A game-based, physical activity coaching application for older adults: design approach and user experience in daily life. Games Health J. **9**, 215–226 (2020).https://doi.org/10.1089/g4h.2018.0163
18. Salazar-Cardona, J., Arango-López, J., Gutiérrez-Vela, F.L.: PL/PX Platform: online tool for the evaluation of fun and game experiences. In: Conference: II Congreso Español de Videojuegos 2023. ceur-ws.org, Madrid (2023)
19. Salazar-Cardona, J.A., Cano, S., Gutiérrez-Vela, F.L., Arango, J.: Designing a tangible user interface (TUI) for the elderly based on their motivations and game elements. Sensors **23**, 9513 (2023). https://doi.org/10.3390/s23239513

Mental Stress Analysis During Visual- And Text-Based Language Learning by Measuring Heart Rate Variability

Katsuyuki Umezawa[1]([✉]), Takumi Koshikawa[1], Makoto Nakazawa[2], and Shigeichi Hirasawa[3]

[1] Shonan Institute of Technology, Kanagawa, Japan
umezawa@info.shonan-it.ac.jp
[2] Junior College of Aizu, Fukushima, Japan
[3] Waseda University, Tokyo, Japan
https://www.shonan-it.ac.jp

Abstract. Visual-based programming languages that allow programming by arranging blocks have become popular as an introductory method to learn programming. In contrast, programming experts generally use text-based programming languages such as C and Java. However, a seamless methodology for transitioning from this visual-based language to a text-based language has not yet been established. Thus, this research aims to establish a methodology that facilitates this transition by bridging the gap between the two languages and clarifying the variations in the biometric information of the learners of both languages. In this study, we measured the heart rate variability of the participants and evaluated the variations in mental stress encountered while learning both visual- and text-based languages.The experimental results confirmed that the heart rate variability value decreased (i.e., stress level increased) during visual language learning. As a conclusion of this study, the rationale became clear that the design of an intermediate type language used by students at various levels, should not cause mental stress to students at any level.

Keywords: visual-based language · text-based language · learning analysis · reart rate variability · mental stress

1 Introduction

In recent years, visual programming languages (hereinafter referred to as visual-based languages) have been employed as an introductory method to teach programming to secondary and higher-secondary school students. At universities or college, the students learn text programming languages such as C and Java (hereinafter referred to as text-based languages). A not insignificant percentage of students get frustrated in the early stages of learning a text-based language. However, a seamless transition method from visual-based language to text-based language has not yet been established.

© The Author(s), under exclusive license to Springer Nature Switzerland AG 2024
Á. Rocha et al. (Eds.): WorldCIST 2024, LNNS 988, pp. 13–22, 2024.
https://doi.org/10.1007/978-3-031-60224-5_2

The current research project aims to establish a methodology that facilitates seamless transitioning from a visual-based language to a text-based language. In addition to applying conventional post-learning evaluation methods, we measured the biological information reflecting the state during learning, such as brain waves, gaze, and facial expressions while learning. Subsequently, we analyzed and evaluated whether the intermediate-type learning contents acts as an intermediary between the visual- and text-based languages and contributes to this transition between languages. This research project will enable novice programmers to start learning programming using visual-based languages, and thereafter, seamlessly and voluntarily transition to build expertise through text-based languages.

This study aims to clarify the differences during learning between visual- and text-based languages by analyzing the biological information during learning. Specifically, we focused on heart rate variability (HRV), which is suitable for evaluating mental stress and the variations in stress during visual- and text-based language learning.

2   Previous Work

2.1   Comparative Study of Visual- and Text-Based Language

Several studies have comparatively analyzed visual- and text-based languages. A previous research [1] reported that in case of complex code structure of a program (e.g., a double loop), visual-based language can resolve misunderstandings to a greater extent. Moreover, visual languages are highly advantageous for beginners. Another study [2] stated that Scratch, one of the visual languages, can learn devise logical thinking as it avoids syntax errors, but it cannot learn to generate programs. [3] compared visual- and text-based languages from a psychological perspective, observing that text language programmers first draw mental representations of the control flow, followed by mental representations of the data flow, whereas visual-based language programmers follow the opposite approach. A previous study [4] investigated the effects of visual- and text-based language programming environments on students' learning outcomes. Nonetheless, the authors could not demonstrate the statistical superiority of visual-based languages in terms of utility and efficiency, highlighting the need for further investigation of hybrid languages.

2.2   Hybrid Languages

Considerable amount of hybrid research has been conducted. A previous study [5] proposed an environment for using a hybrid language of visual and textual languages. They claimed that the proposed environment is suitable for training beginner-level programmers. Moreover, [6] compared text-based languages, visual-based languages, and hybrid languages, demonstrating that hybrid languages exhibit features of both languages and can outperform visual- and text-based languages in certain domains. Previous research [7] used PencilCode to compare the skill effects of migration using visual, hybrid, and text-based languages. Compared to text-based language learners, visual-based language learners could more readily understand loops and variables. In previous research [8], they created tools such as visual-based display, text-based display, and side-by-side view of both visual and text to evaluate the extent of migration. The results indicated that, compared to transitioning from visual- to text-based, transitioning using hybrid programming improved the students' understanding of programming fundamentals, memorization, and ease of transition by more than 30%.

2.3   Gap Between Visual- and Text-Based Languages

Numerous studies have been conducted to explore the differences between visual- and text-based languages. Prior research [9] has indicated a gap between visual- and text-based languages. The authors examined the procedure of migrating from a visual-based language (MIT App Inventor 2) to a text-based language (Android Studio) using the Java Bridge Code Generator as a mediator of knowledge transfer. The study claimed that the Java Bridge Code Generator aided in bridging the gap between visual- and text-based languages. Moreover, [10] experimentally evaluated the differences in knowledge transfer after transitioning to a text-based language between learners who started programming using a visual-based language and those who started with a text-based language. As such, no significant differences were observed between the two groups, because the comparative evaluation was performed after the learners acquired the skills of text-based language rather than evaluating their skills during transitioning.

2.4   Our Previous Research

In our previous studies [11,12], we experimentally confirmed that using intermediate content promotes the comprehension of text-based language, which was proposed to be placed as intermediary module between visual-based language learning and text-based language learning. Through questionnaire, we determined that the proposed intermediate content exhibit features common to visual- and text-based languages [13,14]. More recently, we used electroencephalography (EEG) to monitor the learning progress of learners. We acquired the EEG information from participants engaged in a keyboard typing task, indicating that the value of β/α increases for difficult tasks [15,16]. Upon further analysis of the EEG, a notable difference was observed in the EEG when solving problems in a visual-based programming language (Scratch) and a text-based programming language (C). Specifically, for the visual-based programming language, the value of β/α did not increase with the difficulty of the task, thereby suggesting the existence of various pathways of thinking during the learning process of visual and text-based programming languages [17].

Our previous studies suggested the potential differences in biological responses when learning visual- and text-based languages. Therefore, in this study, we analyzed stress levels by measuring heart rate variability, thereby ascertaining the differences between the two types of language.

3 Experimental Method

3.1 Experiment Participants

The experiment engaged seven fourth-year students from the Shonan Institute of Technology, Japan, all of whom had studied programming-related courses for several years. Ideally, the participants would in the transition stage from visual- to text-based languages. However, for this experiment, fourth-year university students with experience in text-based languages were included. All participants exhibited similar programming skills.

3.2 Web Service Used for Experiments

Google blockly[18] (Fig. 1) was used for programming in a visual-based language. Although this site allows the user to engage in tasks such as puzzles and mazes, this study targeted music tasks to match the contents with the text-based language. In addition, JSFiddle[19] (Fig. 2) was used for programming in a text-based language. This site is an integrated development environment that can execute JavaScript, which is a text-based language. Music can be produced with beep sounds by adding the Beeplay library as a resource setting.

Fig. 1. Screen of Google blockly   **Fig. 2.** Screen of JSFiddle

3.3 Tasks Used for Experiment

The participants were allowed to practice once before performing the actual experiment twice. Specifically, participants were assigned a musical score and instructed to write a program that produces the sounds according to the musical score. The songs targeted in each exercise and experiment are listed in Table 1. These scores were printed out and presented to the participants during the experiment. In particular, the musical notes were represented according to international notation (C3, C4, etc.) below the notes in the music score. Specifically, the participants were asked to create a program that reproduces the music outlined in Table 1, using the provided musical score. If there were utilizing a visual-based language, they were asked to arrange blocks using the mouse and, if they were using a text-based language, they were instructed to enter the code using the keyboard.

Table 1. Experimental song

experiment	song title
practice	Froggy's Song
experiment 1	Mary Had A Little Lamb
experiment 2	Jingle Bells

3.4   Equipment Used in Experiments

Heart Rate Variability Monitor. Generally, the heart rate interval (RR interval) fluctuates and these can be observed even when during resting periods. This periodic variation in the heart rate interval is called "heart rate variability (HRV)." It is generally accepted that when stress levels increase, HRV decreases, while balanced stress levels result in increased HRV.

An Apple "Apple Watch 8" was used to measure HRV. In addition, "HRV analysis/electrocardiogram analyzer" by WMS, Inc. was used as a measurement application. This application can connect with Apple Watch and calculate HRV (root–mean–square of successive differences: rMSSD). When measuring HRV with Apple Watch 8, a finger of the other hand must be placed on the crown part of the watch while wearing it on the wrist for approximately 30 s.

3.5   Experiment Flow

The overall flow of the experiment is illustrated in Fig. 3, wherein the participants were randomly segmented into a group that started programming with a text-based language (Group A) and a group that started with a visual-based language (Group B). As discussed in the 3.4 Sect., HRV cannot be measured during the experiment (during programming) as it requires to touch the crown. Thus, we decided to measure the HRV for 30 s before and after the experiment, as depicted in Fig. 3. In addition, the input information by the key logger was recorded during each experiment.

4   Experimental Result

The HRV of Group A (participants A to D) is plotted in Fig. 4, whereas that of Group B (participants E to G) are portrayed in Fig. 5. To address concerns regarding the potential influence of experiment order on changes in HRV, the participants were randomly divided into Group A and Group B for this study,

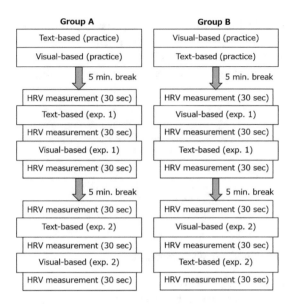

Fig. 3. Experiment flow

with the order of the experiments being switched. The figures reveal that, except for participants C and G, HRV values exhibited an increase before and after the text-based language experiment in both groups. Additionally, for the visual-based language experiment, participants in both groups, except for participant B, experienced a decrease in HRV before and after the experiment. Generally, HRV is known to decrease as stress levels rise, while it increases when stress balance is maintained. Thus, several participants were able to maintain their stress balance with the text-based language, whereas it appears that the visual-based language induced elevated stress levels for most participants. The statistical validity of these findings are investigated in the subsequent sections.

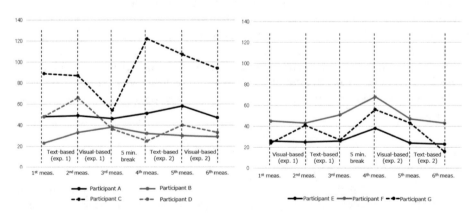

Fig. 4. HRV in Group A   **Fig. 5.** HRV in Group B

5   Analysis

As depicted in Figs. 4 and 5 from the previous section, the trend of changes in HRV before and after the visual-based language and text-based language experiments appears to be consistent regardless of the experiment order. Consequently, in the subsequent analysis within this section, the measurements before and after the text-based language (exp.1) experiment for Group A (i.e., the first and second measurements) and the measurements before and after the text-based language (exp.1) experiment for Group B (i.e., the second and third measurements) are treated as identical. Similarly, measurements before and after the same type of experiments in both Group A and Group B are considered to be equivalent. The rationale behind this data handling approach is elaborated in Sect. 5.1.

Furthermore, according to the findings reported in [20], the distribution of rMSSD does not follow a normal distribution. However, the rMSSD values conformed to a normal distribution through natural logarithmic transformation. Moreover, previous research [21] states that parametric tests, such as the t-test, can be employed irrespective of the sample size (where $n \geq 2$), assuming the independence of observations and the normality and homoscedasticity of the population distribution. Additionally, the paired t-test does not necessitate the assumption of homoscedasticity. Therefore, in the forthcoming analysis, a paired t-test will be conducted using the rMSSD data transformed by natural logarithm.

In the measurement of HRV indices, such as rMSSD, using ultrashort time measurements of 30 s as employed in this experiment, a comparison of results across experiments with varying measurement times within the same group should precede the statistical testing between groups, as highlighted in a previous study [20]. However, owing to the limited measurement period of only 30 s, tests within the same group were omitted in this analysis.

5.1   Test of Difference Between Group A and Group B

To ascertain whether there is a difference in the measurement results before and after the same type of experiment for Groups A and B, we perform a Mann-Whitney U-test, one of the nonparametric tests of difference between two groups with no correspondence. We will compare the difference before and after the text type of group A (exp.1 and exp.2) (8 data in total) and the difference before and after the text type of group B (exp.1 and exp.2) (6 data in total). The same test is performed for the visual-based language. The results are shown in Table 2. From the Mann-Whitney test table, the significance level is $\alpha = 0.05$, and from the two-sided 5% test table, the limit value is 8.

Table 2. Mann-Whitney U-test result

experiment	test statistic	limit value
text-based	13.5	8
visual-based	21	8

As observed in Table 2, both the text-based and visual-based language experiments yielded a "test statistic ≥ limit value" outcome, indicating that no significant difference can be inferred between the data of the two groups. Consequently, the measurements before and after the same type of experiment in both Groups A and B were treated as equivalent.

5.2   Test for Population Mean Difference

Furthermore, we conducted paired t-tests to assess whether there is a difference in the mean values of the natural logarithm-transformed HRV (rMSSD) before and after each text-based (exp.1), visual-based (exp.1), text-based (exp.2), and visual-based (exp.2) experiment, considering all participants. The test results are presented in Table 3 and depicted in Fig. 6.

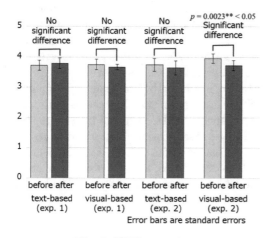

Fig. 6. HRV mean test

Table 3. HRV mean test

exp.	b/a	mean	p-value
text	before	3.7229	0.2604
(exp.1)	after	3.7899	
visual	before	3.7552	0.3005
(exp.1)	after	3.6761	
text	before	3.7443	0.2811
(exp.2)	after	3.6418	
visual	before	3.9580	0.0023**
(exp.2)	after	3.7209	

As listed in Table 3 and Fig. 6, a significant decrease in HRV values was observed before and after the visual-based language experiment (exp.2). However, no statistically significant differences were observed in the remaining experiments. These findings suggest that text-based language learning did not have an impact on stress balance, while visual-based language learning resulted in an increase in stress levels.

In addition to the above analysis, we also attempted verification employing Wilcoxon signed-rank sum test, which is a type of nonparametric test. The conclusion was exactly the same as the above t-test.

6   Conclusion and Future Work

In our study, we measured the HRV of learners to investigate potential differences in mental stress during the learning of visual- and text-based languages. Our findings revealed a decrease in HRV (rMSSD) during visual language learning, suggesting that stress levels are higher when learning a visual-based language compared to a text-based one. This result is likely reflective of the fact that the participants were students not transitioning from visual- to text-based languages, but were fourth-year university students with a reasonable understanding of text-based language.

In future, we plan to develop an intermediate language that bridges the gap between visual- and text-based languages. Based on the conclusions of this research, an intermediate language should be developed such that it does not impose stress on students with various levels of understanding and proficiency.

Acknowledgments. Part of the work reported here was conducted as a part of the research project "Research on e-learning for next-generation" of Waseda Research Institute for Science and Engineering, Waseda University. Part of this work was supported by JSPS KAKENHI Grant Numbers JP22H01055, JP21K18535, and JP20K03082. Research leading to this paper was partially supported by the grant of "ICT and Education" of JASMIN.

Research Ethics. All experiments were approved by the Research Ethics Committee of Shonan Institute of Technology . We received written informed consent from the participants and their parents or guardians.

References

1. Mladenović, M., Boljat, I., Žanko, Ž: Comparing loops misconceptions in block-based and text-based programming languages at the k-12 level. Educ. Inf. Technol. **23**(4), 1483–1500 (2018)
2. Robinson, W.: From scratch to patch: Easing the blockstext transition. In: In Proceedings of the 11th Workshop in Primary and Secondary Computing Education (ACM), pp. 96–99 (2016)
3. Navarro-Prieto, R., Canas, J.J.: Are visual programming languages better? The role of imagery in program comprehension. Int. J. Hum.-Comput. Stud. **54**(6), 799–829 (2001)
4. Xu, Z., Ritzhaupt, A.D., Tian, F., Umapathy, K.: Block-based versus text-based programming environments on novice student learning outcomes: a meta-analysis study. Comput. Sci. Educ. **29**, 177–204 (2001)
5. Daskalov, R.: Pashev, George, Gaftandzhieva, Silvia: hybrid visual programming language environment for programming training. TEM Journal **10**, 981–986 (2021)
6. Weintrop, D., Wilensky, U.: Between a block and a typeface: Designing and evaluating hybrid programming environments. In: IDC 2017: Proceedings of the 2017 Conference on Interaction Design and Children, pp. 183–192 (2017)
7. Weintrop, D.: Blocks, text, and the space between: the role of representations in novice programming environments. In: In 2015 IEEE Symposium on Visual Languages and Human-Centric Computing (VL/HCC), pp. 301–302 (2015)
8. Alrubaye, H., Ludi, S., Mkaouer, M.W.: Comparison of block-based and hybrid-based environments in transferring programming skills to text-based environments. In: Proceedings of the 29th Annual International Conference on Computer Science and Software Engineering, pp. 100–109 (2019)

9. Tóth, T., Lovászová, G.: Mediation of knowledge transfer in the transition from visual to textual programming. Inform. Educ. **20**, 489–511 (2021)

10. Weintrop, D., Wilensky, U.: Transitioning from introductory block-based and text-based environments to professional programming languages in high school computer science classrooms. Comput. Educ. **142**, 103646 (2019)

11. Umezawa, K., Ishida, K., Nakazawa, M., Hirasawa, S.: A proposal and evaluation of intermediate content for transition from visual to text-based languages, pp. 1–7. The Institute of Electronics, Information and Communication Engineers (IEICE) Technical report (2022)

12. Katsuyuki U., Kouta I., Makoto N., Shigeichi H.: Proposal and evaluation of intermediate content for the transition from visual to text-based programming languages. In: Proceedings of the 56th Hawaii International Conference on System Sciences (HICSS 2023), pp. 83–92 (2023)

13. Katsuyuki, U., Makoto, N., Shigeichi, H.: Seamless transition from visual-type to text-type languages. In: Proceedings of the 84th National Convention of IPSJ, vol. 4, pp. 519–520 (2022)

14. Katsuyuki U., Makoto N., Shigeichi H.: A proposal for intermediate content: transition from visual to text-based languages. In: Proceeding of the World Conference on Computers in Education (WCCE2022), p. 1 (2022)

15. Umezawa, Katsuyuki, Saito, Tomohiko, Ishida, Takashi, Nakazawa, Makoto, Hirasawa, Shigeichi: Learning state estimation method by browsing history and brain waves during programming language learning. In: Rocha, Álvaro., Adeli, Hojjat, Reis, Luís Paulo., Costanzo, Sandra (eds.) WorldCIST'18 2018. AISC, vol. 746, pp. 1307–1316. Springer, Cham (2018). https://doi.org/10.1007/978-3-319-77712-2_125

16. Umezawa, K., Saito, T., Ishida, T., Nakazawa, M., Hirasawa, S.: Learning-state-estimation method using browsing history and electroencephalogram in e-learning of programming language and its evaluation. In: Proceeding of the International Workshop on Higher Education Learning Methodologies and Technologies Online (HELMeTO 2020), pp. 22–25 (2020)

17. Umezawa, K., Ishii, Y., Nakazawa, M., Nakano, M., Kobayashi, M., Hirasawa, S.: Comparison experiment of learning state between visual programming language and text programming language. In: 2021 IEEE International Conference on Engineering, Technology & Education (TALE), pp. 01–05 (2021)

18. Blockly games. https://blockly.games/. Accessed 17 January 2023

19. Jsfiddle. https://jsfiddle.net/. Accessed 17 January 2023

20. Pecchia, L., Castaldo, R., Montesinos, L., Melillo, P.: Are ultra-short heart rate variability features good surrogates of short-term ones? State-of-the-art review and recommendations. Healthc. Technol. Lett. **5**, 94–100 (2018)

21. Mizumoto, A.: A comparison of statistical tests for a small sample size: application to corpus linguistics and foreign language education and research. The Institute of Statistical Mathematics cooperative research report **238**, 1–14 (2010)

The Challenges of Learning Assessment in the Age of Artificial Intelligence

Bruno F. Gonçalves[1]([⊠]) [iD], Maria Raquel Patrício[1] [iD], and Amália Comiche[2]

[1] CIEB, Polytechnic Institute of Bragança, Bragança, Portugal
bruno.goncalves@ipb.pt
[2] Catholic University of Mozambique, Beira, Mozambique

Abstract. The continuous growth of artificial intelligence in the world and the consequent integration of this type of technology into the various economic sectors of societies seems to be triggering new relationships with machines, but also new ways of working. Education, as one of the most important sectors of a nation, can no longer escape this new reality of integrating this type of technology into teaching and learning processes. It is precisely because of this fact that this research emerges, which aims not to set aside or exclude this (r)evolution, but rather to consider them as important and useful tools for innovation in education, putting them at the service of this sector. In this sense, we believe that it is essential to carry out a study to identify and understand the main challenges in carrying out the process of assessing students in the age of artificial intelligence. A systematic review of the literature will be carried out, focusing only on the three years - the artificial intelligence boom - and through it we will try to identify the major challenges that educational agents, but especially teachers, face in assessing students, contributing to literacy in the area, but also to a serious debate on this issue that is already being discussed so much in educational institutions. The results suggest that there is a set of challenges that teachers have to deal with, which, according to the content analysis carried out, are related to authenticity, ethics and fraud.

Keyword: Artificial intelligence · Digital technologies · Education · Learning assessment

1 Introduction

Over the last year, much has been discussed about the impact of artificial intelligence on education, its advantages, disadvantages, challenges, opportunities and impacts associated with integrating this new era of technology - web 4.0 - into education and other teaching and learning processes. The countless studies that have emerged in this area have made a significant and indispensable contribution to a broader understanding of how AI can and should be integrated into education. Reflection and debate are very present in both the educational and scientific communities and are undoubtedly absolutely useful, relevant and central to the current reality of schools. We can no longer have digitally native students in a school that still teaches as it did in the last century.

Á. Rocha et al. (Eds.): WorldCIST 2024, LNNS 988, pp. 23–32, 2024.
https://doi.org/10.1007/978-3-031-60224-5_3

Teaching has to be reinvented, all of it, without exception. From updating the subjects to be taught, the process of teaching and transmitting knowledge, to lesson models and also with regard to the process of assessing students, whether in a face-to-face classroom context or in a virtual classroom context. Change is essential and urgent, otherwise the school will lose its main function: to educate!

It is in this sense that a serious reflection on the student assessment process is essential, especially at a time when artificial intelligence is present in the lives of citizens and in the lives of organizations, becoming a central aspect of the life of educational institutions, regardless of the cycle of studies.

Thus, this research seeks to identify the challenges of the student assessment process in the age of artificial intelligence, so that educational agents, namely heads of educational institutions and teachers, reflect in their educational communities on the best ways to assess today, given the exponential growth of web 4.0 technologies.

2 Methodology

Through this research we seek to make a systematic review of the literature identifying the main challenges of learning assessment in the age of artificial intelligence. For this, the systematic literature review is adopted as the investigative methodology to support the study. The systematic review of the literature will be carried out with the support of a pre-defined set of criteria that will be essential for the serration of information on the theme addressed, namely: (i) Time interval: 2019–2023; (ii) Documents: reference articles; (iii) Search language: English; (iv) Bibliometric databases: Scopus; (v) Other databases: Google Scholar; (vi) Keywords: challenges of learning assessment in the age of artificial intelligence; challenges that artificial intelligence brings to student assessment; artificial intelligence and the assessment of student learning; artificial intelligence and student assessment. It should be noted that although AI emerged several decades ago, we considered the 2019–2023 timeframe due to the fact that it was the AI boom and therefore there was more and more current literature.

The following set of exclusion criteria were also considered: (i) Articles published outside the established time frame; (ii) Documents that are not reference articles; (iii) Articles in languages other than English; (iv) Duplicate articles in different databases.

Based on the criteria previously established, the general framework of the documents found is presented (Table 1):

Table 1. General table of documents found

ID	Article name	Year	Authors
1	Implementation of Artificial, Intelligence in Imparting Education and Evaluatings Student	2019	Kumar, D. N.[1]
2	Vision, challenges, roles and research issues of Artificial Intelligence in Education	2020	Hwang, G. J., Xie, H., Wah, B. W., & Gašević, D. [2]

(continued)

Table 1. (*continued*)

ID	Article name	Year	Authors
3	Artificial Intelligence for Student Assessment: A Systematic Review	2021	González-Calatayud, V., Prendes-Espinosa, P., & Roig-Vila, R.[3]
4	A Review of Artificial Intelligence (AI) in Education from 2010 to 2020	2021	Zhai, X., Chu, X., Chai, C. S., Jong, M. S. Y., Istenic, A., Spector, M., … & Li, Y. [4]
5	A Review on Artificial Intelligence in Education	2021	Huang, J., Saleh, S., & Liu, Y. [5]
6	Artificial Intelligence for Assessment and Feedback to Enhance Student Success in Higher Education	2022	Hooda, M., Rana, C., Dahiya, O., Rizwan, A., & Hossain, M. S. [6]
7	The Promises and Challenges of Artificial Intelligence for Teachers: a Systematic Review of Research	2022	Celik, I., Dindar, M., Muukkonen, H., & Järvelä, S. [7]
8	Assessment in the age of artificial intelligence	2022	Zachari Swiecki, Hassan Khosravi, Guanliang Chen, Roberto Martinez-Maldonado, Jason M. Lodge, Sandra Milligan, Neil Selwyn, Dragan Gaˇsevi´c [8]
9	Artificial Intelligence Applications in K-12 Education: A Systematic Literature Review	2022	Zafari, M., Bazargani, J. S., Sadeghi-Niaraki, A., & Choi, S. M. [9]
10	Trends of Artificial Intelligence for Online Exams in Education	2022	Babitha, M. M., Sushma, C., & Gudivada, V. K. [10]
11	Systematic literature review on opportunities, challenges, and future research recommendations of artificial intelligence in education	2022	Xia, Q., Chiu, T. K., Zhou, X., Chai, C. S., & Cheng, M. [11]
12	Ethical principles for artificial intelligence in education	2023	Nguyen, A., Ngo, H. N., Hong, Y., Dang, B., & Nguyen, B. P. T. [12]
13	Establishing a delicate balance in the relationship between artificial intelligence and authentic assessment in student learning	2023	Lawrie, G. [13]
14	Exploring the Use of ChatGPT as a Tool for Learning and Assessment in Undergraduate Computer Science Curriculum: Opportunities and Challenges	2023	Qureshi, B. [14]

(*continued*)

Table 1. (*continued*)

ID	Article name	Year	Authors
15	Machine Learning and Deep Learning in Assessment	2023	Jiao, H., He, Q., & Yao, L. [15]
16	Artificial intelligence in education	2023	Holmes, W., Bialik, M., & Fadel, C. [16]
17	Can we and should we use artificial intelligence for formative assessment in science?	2023	Li, T., Reigh, E., He, P., & Adah Miller, E. [17]
18	A critical evaluation, challenges, and future perspectives of using artificial intelligence and emerging technologies in smart classrooms	2023	Dimitriadou, E., & Lanitis, A. [18]
19	Education in the Era of Generative Artificial Intelligence (AI): Understanding the Potential Benefits of ChatGPT in Promoting Teaching and Learning	2023	Baidoo-Anu, D., & Ansah, L. O. [19]
20	The Use of Artificial Intelligence (AI) in Online Learning and Distance Education Processes: A Systematic Review of Empirical Studies	2023	Dogan, M. E., Goru Dogan, T., & Bozkurt, A. [20]
21	Empowering learners for the age of artificial intelligence	2023	Gašević, D., Siemens, G., & Sadiq, S. [21]

Four documents were excluded because they were outside the lines of the research and, in this sense, twenty-one documents were considered for the purposes of this research.

The data collected from the documents that emerged from the search were categorized and treated in Microsoft Excel. The content analysis of the selected documents showed that there are a number of challenges in evaluation, which will be presented and discussed in the following section.

3   Challenges of Learning Assessment

Several documents were found that address some of the challenges of learning assessment in the age of artificial intelligence, so below we discuss the main results obtained.

– The use of AI to improve and automate assessment results in more effective classification [22–26]. However, most automatic classification is homogeneous and applied to only a few disciplines and domains, which indicates that the application of AI is at an early stage of development. In this sense, migrating the technology to authentic educational environments would present enormous challenges [26, 27];

- AI seems to help predict student performance, particularly in e-learning [28–30], such as the quality of participation, the completion of activities, among other aspects. Massive Open Online Courses (MOOCs) are clearly an example of the integration of this type of technology. However, selecting data for prediction is a challenge [11]. [29] argued that student data used in classical statistics may not be suitable for AI predictive models. In other words, selecting suitable data for predictive models of student performance remains a challenge, as the data is not the same as that used in traditional educational research [11];
- Assessing the impacts of AI-supported learning design on students' performances and perceptions can be another challenge. Other aspects can be taken into account, such as students' learning performance, learning motivation, learning anxiety, self-efficacy, cognitive load, the impacts of AI-supported learning designs on the performance and experience of students with different personal characteristics [2];
- The AI assessment model can also present other challenges, such as the "marginalization of professional knowledge", the "black box" of responsibility by placing decisions in the hands of programmers, the restriction of the role that pedagogy plays in assessment, the limitation of responsibility and the scope of learning and the pedagogy of surveillance [21];
- Research shows that a machine cannot take on the role of a teacher, and the way AI works and carries out processes in the context of teaching is far removed from human intelligence [31] and partly due to the lack of transparency in decision-making algorithms [32]. In this sense, AI training for teachers and students is important. It is essential to train teachers to use this technology [33], but not just on the basis of learning the tools, but also on the basis of pedagogical reference models that give meaning to the development of lessons. Several authors [34–36] have determined that AI requires specific training for students as future professionals, as they need to understand the characteristics, possibilities and limitations of these intelligent systems;
- Also other investigative work [16] raises some questions about the collection of accurate and comprehensive information about the student, the possibility of cheating and subversion of the system, and the dependence on efficient algorithms and large amounts of personal data, which raises ethical and of privacy;
- Another challenge is to ensure equity in the application of AI in education [5]. With the development of AI, developing countries run the risk of exacerbating divisions in education through new technologies. Since most AI algorithms come from developed countries, they cannot fully take into account the conditions in developing countries and cannot be applied directly [37];
- Ethical and security issues are another major challenge arising from the collection, use and dissemination of data. AI has raised many ethical questions in terms of providing personalized advice to students, collecting personal data, data privacy, and ownership of responsibilities and data-feeding algorithms [12, 38, 39]. Strengthening oversight of AI technology and its products requires the public to discuss the ethics, responsibility and safety involved [5];

- The creators of AI teaching products must also understand the way teachers work and create a plan for using teaching products that is convenient for teachers [5]. Only in this way will it be possible to ensure that teachers acquire skills in AI technologies for student assessment;
- Another challenge is the limited technical capacity of AI [7] as, for example, the AI may not be efficient for scoring graphics or figures and text. [40] reported that an AI-based system failed to assess the complexity of texts when they included images. That is, automated writing evaluation technologies that use AI algorithms have to be improved to provide trustworthy evaluations for teachers [41]. AI-based scoring may sometimes improperly evaluate performance [42];
- The literature shows that AI-generated text detectors are not effective with current sophisticated natural processing language models [43, 44]. This means that, for the purposes of evaluating work carried out by students, teachers must be aware of this situation in order to guarantee a more serious and rigorous evaluation;
- In another research [17], three more challenges related to evaluation were identified. Thus, AI can: (i) Be biased and not take into account the cultural and linguistic diversity of students; (ii) Not being able to recognize and value students' most unexpected and rich ideas; (iii) Represent risks to the art and practice of teaching, with potential negative consequences for student learning;
- The lack of digital competences of teachers [45] and the lack of technical infrastructure in schools [46] are two more challenges in integrating AI into education;
- Finally, other important challenges in applying AI in assessment are related to the interpretability and validity of results obtained through machine learning and deep learning [15].

The assessment process in a face-to-face or virtual teaching-learning context is, without a doubt, a subject that has received a lot of attention from the educational community. The literature shows that teachers have limited capacity and skills to engage in high-quality assessment practices that boost learning [19]. Perhaps for this reason, educators have consistently called on teachers to develop skills to engage students in high-quality assessment practices [47–49]. Training and empowering teachers in this area is essential to develop the skills needed to harness the power of AI to engage in high-quality assessment practices that improve student learning [19].

4   Conclusions

The research made it possible to contribute to the discussion of artificial intelligence in education, especially with regard to the main challenges in the assessment process in the new AI age.

According to the research, there are several challenges facing the school as an institution and teachers as players in this evaluation process. These challenges emerge from the evolution of the times, the transformation of the school as a constantly changing context and also from the change and innovation of teaching practices in the teaching-learning processes.

The challenges set out in the results are related to the lack of skills and knowledge in artificial intelligence, conditioning their literacy in the area, a situation that could jeopardize the integration of AI into students' assessment processes. Another challenge seems to be the loss of teachers' roles in assessment processes, which could consequently lead to the teacher's role becoming less visible, whether in a face-to-face classroom context or online. The adequacy of teachers' assessment methods is another challenge, since these methods must be adjusted to the integration of AI in education. Understanding the results of AI is another challenge that must be taken into account, since it is up to the teacher and not AI to make decisions about the results. Another challenge identified in the results is the prevention and detection of plagiarism, which can be practiced through AI tools that continue to emerge on the market. It is also crucial for teachers to have digital AI skills in order to understand whether the results generated by AI are actually equitable. Not only are digital skills important, but also pedagogical skills, essentially best practices in the use of AI in both the teaching process and student assessment.

The big challenge is not to limit the use of AI in education, to set it aside or avoid it, but to gradually integrate it into the teaching-learning process, in teaching, in the proposal of activities and in student assessment, among other dimensions. The challenge is therefore to integrate AI into education! However, it is necessary to train teachers in the proper and contextualized use of AI technologies in the teaching-learning process, but also the students who are key players in this process. In addition to AI training aimed at acquiring skills, it is also important to collaborate and share with colleagues, to promote students' digital literacy and to promote sessions on privacy, safety and ethics.

AI has made a contribution to education in all its dimensions, however, it is necessary to train educational agents in its correct use, making them especially aware of ethical issues. We are in a new world in which all educational agents have to adapt and processes have to be adapted to this new reality. It's undoubtedly a great opportunity to innovate in education with everyone!

5   Limitations and Future Research

There are two limitations to the study that it is important to mention, firstly the fact that the subject of artificial intelligence is very recent, which on the one hand makes it difficult to find sources on the subject of the research, and on the other hand it is not yet possible to have access to consolidated literature in the area. Another limitation is that the systematic literature review was not comprehensive or representative of all the evaluation criteria for education research, as it was limited to the keywords defined in the methodology section.

Despite these limitations, our study provides valuable insights into the challenges of assessment in learning contexts where artificial intelligence is increasingly present.

As future work, we believe it is important to identify and characterize a set of intelligence tools that will enable teachers to guarantee greater efficiency and accuracy in the student assessment process.

Acknowledgment. This work has been supported by FCT – Fundação para a Ciência e Tecnologia within the Project Scope: UIDB/05777/2020.

References

1. Kumar, D.N.M.: Implementation of artificial intelligence in imparting education and evaluating student performance. J. Artif. Intell. Capsul. Netw. **1**(1), 1–9 (2019)
2. Hwang, G.-J., Xie, H., Wah, B.W., Gašević, D.: Vision, challenges, roles and research issues of artificial intelligence in education. Comput. Educ. Artif. Intell. **1**, 100001 (2020). Elsevier
3. González-Calatayud, V., Prendes-Espinosa, P., Roig-Vila, R.: Artificial intelligence for student assessment: a systematic review. Appl. Sci. **11**(12), 5467 (2021)
4. Zhai, X., et al.: A review of artificial intelligence (AI) in education from 2010 to 2020. Complexity **2021**, 1–18 (2021)
5. Huang, J., Saleh, S., Liu, Y.: A review on artificial intelligence in education. Acad. J. Interdiscip. Stud. **10**(3), 206 (2021)
6. Hooda, M., Rana, C., Dahiya, O., Rizwan, A., Hossain, M.S.: Artificial intelligence for assessment and feedback to enhance student success in higher education. Math. Probl. Eng. **2022**, 1–19 (2022)
7. Celik, I., Dindar, M., Muukkonen, H., Järvelä, S.: The promises and challenges of artificial intelligence for teachers: a systematic review of research. TechTrends **66**(4), 616–630 (2022)
8. Swiecki, Z., et al.: Assessment in the age of artificial intelligence. Comput. Educ. Artif. Intell. **3**, 100075 (2022)
9. Zafari, M., Bazargani, J.S., Sadeghi-Niaraki, A., Choi, S.-M.: Artificial intelligence applications in K-12 education: a systematic literature review. IEEE Access **10**, 61905–61921 (2022)
10. Babitha, M.M., Sushma, C., Gudivada, V.K.: Trends of Artificial Intelligence for online exams in education. Int. J. Early Child. Spec. Educ. **14**(01), 2457–2463 (2022)
11. Xia, Q., Chiu, T.K.F., Zhou, X., Chai, C.S., Cheng, M.: Systematic literature review on opportunities, challenges, and future research recommendations of artificial intelligence in education. Comput. Educ. Artif. Intell. **4**, 100118 (2022)
12. Nguyen, A., Ngo, H.N., Hong, Y., Dang, B., Nguyen, B.-P.T.: Ethical principles for artificial intelligence in education. Educ. Inf. Technol. **28**(4), 4221–4241 (2023)
13. Lawrie, G.: Establishing a delicate balance in the relationship between artificial intelligence and authentic assessment in student learning. Chem. Educ. Res. Pract. **24**(2), 392–393 (2023)
14. Qureshi, B.: Exploring the use of ChatGPT as a tool for learning and assessment in undergraduate computer science curriculum: Opportunities and challenges. arXiv Prepr. arXiv2304.11214 (2023)
15. Jiao, H., He, Q., Yao, L.: Machine learning and deep learning in assessment. Psychol. Test Assess. Model. **64**(1), 178–189 (2023)
16. Holmes, W., Bialik, M., Fadel, C.: Artificial Intelligence in Education. Globethics Publications (2023)
17. Li, T., Reigh, E., He, P., Adah Miller, E.: Can we and should we use artificial intelligence for formative assessment in science?. J. Res. Sci. Teach. **60**(6), 1–5 (2023)
18. Dimitriadou, E., Lanitis, A.: A critical evaluation, challenges, and future perspectives of using artificial intelligence and emerging technologies in smart classrooms. Smart Learn. Environ. **10**(1), 1–26 (2023)
19. Baidoo-Anu, D., Owusu Ansah, L.: Education in the Era of Generative Artificial Intelligence (AI): Understanding the Potential Benefits of ChatGPT in Promoting Teaching and Learning. Available SSRN 4337484 (2023)
20. Dogan, M.E., Goru Dogan, T., Bozkurt, A.: The use of artificial intelligence (AI) in online learning and distance education processes: a systematic review of empirical studies. Appl. Sci. **13**(5), 3056 (2023)

21. Gašević, D., Siemens, G., Sadiq, S.: Empowering learners for the age of artificial intelligence. Comput. Educ. Artif. Intell. **4**, 100130 (2023)
22. Çebi, A., Karal, H.: An application of fuzzy analytic hierarchy process (FAHP) for evaluating students' project. Educ. Res. Rev. **12**(3), 120–132 (2017)
23. Alghamdi, A., Alanezi, M., Khan, F.: Design and implementation of a computer aided intelligent examination system. Int. J. Emerg. Technol. Learn. **15**(1), 30–44 (2020)
24. Fu, S., Gu, H., Yang, B.: The affordances of AI-enabled automatic scoring applications on learners' continuous learning intention: an empirical study in China. Br. J. Educ. Technol. **51**(5), 1674–1692 (2020)
25. Kumar, V., Boulanger, D.: Explainable automated essay scoring: deep learning really has pedagogical value. In: Frontiers in Education, vol. 5, pp. 572367 (2020)
26. Ma, H., Slater, T.: Using the developmental path of cause to bridge the gap between AWE scores and writing teachers' evaluations (2015)
27. Sun, Y.: Application of artificial intelligence in the cultivation of art design professionals. Int. J. Emerg. Technol. Learn. **16**(8), 221–237 (2021)
28. Akmeşe, Ö.F., Kör, H., Erbay, H.: Use of machine learning techniques for the forecast of student achievement in higher education. Inf. Technol. Learn. Tools **82**(2), 297–311 (2021)
29. Costa-Mendes, R., Oliveira, T., Castelli, M., Cruz-Jesus, F.: A machine learning approximation of the 2015 Portuguese high school student grades: a hybrid approach. Educ. Inf. Technol. **26**(2), 1527–1547 (2021)
30. Yu, J.: Academic performance prediction method of online education using random forest algorithm and artificial intelligence methods. Int. J. Emerg. Technol. Learn. **15**(5), 45–57 (2021)
31. Cope, B., Kalantzis, M., Searsmith, D.: Artificial intelligence for education: knowledge and its assessment in AI-enabled learning ecologies. Educ. Philos. Theory **53**(12), 1229–1245 (2021)
32. Espinosa, M.P.P., Cartagena, F.C.: Tecnologías avanzadas para afrontar el reto de la innovación educativa. RIED. Rev. Iberoam. Educ. a Distancia **24**(1), 33–53 (2021)
33. Cabero-Almenara, J., Romero-Tena, R., Palacios-Rodríguez, A.: Evaluation of teacher digital competence frameworks through expert judgement: The use of the expert competence coefficient. J. New Approaches Educ. Res. (NAER Journal) **9**(2), 275–293 (2020)
34. (Hans) Korteling, J.E., van de Boer-Visschedijk, G.C., Blankendaal, R.A.M., Boonekamp, R.C., Eikelboom, A.R.: Human-versus artificial intelligence. Front. Artif. Intell. **4**, 622364 (2021)
35. De Beijing, C.: Consenso de Beijinng sobre a inteligência artificial e a educação (2019). https://en.unesco.org/themes/ict-education
36. Collazos, C.A., Gutiérrez, F.L., Gallardo, J., Ortega, M., Fardoun, H.M., Molina, A.I.: Descriptive theory of awareness for groupware development. J. Ambient. Intell. Humaniz. Comput. **10**, 4789–4818 (2019)
37. Yu, P.K.: The algorithmic divide and equality in the age of artificial intelligence. Fla. L. Rev. **72**, 331 (2020)
38. Bodo, B., et al.: Tackling the algorithmic control crisis-the technical, legal, and ethical challenges of research into algorithmic agents. Yale JL Tech. **19**, 133 (2017)
39. Southgate, E.: Artificial intelligence, ethics, equity and higher education: A 'beginning-of-the-discussion' paper (2020)
40. Fitzgerald, J., et al.: Important text characteristics for early-grades text complexity. J. Educ. Psychol. **107**(1), 4 (2015)
41. Qian, L., Zhao, Y., Cheng, Y.: Evaluating China's automated essay scoring system iWrite. J. Educ. Comput. Res. **58**(4), 771–790 (2020)
42. Lu, X.: An empirical study on the artificial intelligence writing evaluation system in China CET. Big data **7**(2), 121–129 (2019)

43. Williams, C.: Hype, or the future of learning and teaching? 3 Limits to AI's ability to write student essays (2023)
44. Tate, T., Doroudi, S. Ritchie, D., Xu, Y.: Educational research and AI-generated writing: Confronting the coming tsunami (2023)
45. Chiu, T.K.F., Chai, C.: Sustainable curriculum planning for artificial intelligence education: a self-determination theory perspective. Sustainability **12**(14), 5568 (2020)
46. McCarthy, T., Rosenblum, L.P., Johnson, B.G., Dittel, J., Kearns, D.M.: An artificial intelligence tutor: a supplementary tool for teaching and practicing braille. J. Vis. Impair. Blind. **110**(5), 309–322 (2016)
47. Earl, L.M.: Assessment as Learning: Using Classroom Assessment to Maximize Student Learning. Corwin Press, Thousand Oaks (2012)
48. Wiliam, D.: What is assessment for learning? Stud. Educ. Eval. **37**(1), 3–14 (2011)
49. Willis, J., Adie, L., Klenowski, V.: Conceptualising teachers' assessment literacies in an era of curriculum and assessment reform. Aust. Educ. Res. **40**, 241–256 (2013)

OqKay: A Semi-automatic System Approach to Question Extraction

Kayque Lucas Santana dos Santos[1] ⓘ, Aluisio José Pereira[1] ⓘ,
Leandro Marques Queiros[1] ⓘ, Carlos José Pereira da Silva[1] ⓘ,
Alex Sandro Gomes[1] ⓘ, and Fernando Moreira[2(✉)] ⓘ

[1] Centro de Informática, Universidade Federal de Pernambuco, Recife, Brazil
{klss,ajp3,lmq,cjps,asg}@cin.ufpe.br
[2] REMIT, Universidade Portucalense, Porto and IEETA, Universidade de Aveiro, Aveiro,
Portugal
fmoreira@upt.pt

Abstract. Basic education teachers carry out routine activities to develop evaluative exercises. In High School, this task must be aligned with entrance exams. Such practice generally demands recurrent research and intensive question selection. In this sense, this study presents the design of a semi-automatic system approach for extracting Enem questions available in internet repositories. The methodology employed web scraping methods to investigate educational materials, gather evidence of contest editions, employ regular expressions to examine the content and organize the data found within the documents. It also stored a repository of questions for utilization in the proposed system. When evaluating the approach, we obtained an acceptance and performance rate of 68.9% of questions extracted assertively. Thus, automating part of the specialized activity required to extract evaluative questions from online repositories is possible.

Keywords: web scraping · question extraction · Enem · evaluation

1 Introduction

Planning and preparing teaching tools, especially assessment exercises, are an integral and continuous part of the teacher's work. These activities require additional hours during planning, mainly outside of lessons. According to the teaching statute of the state of Pernambuco, Law 11.329 of January 16, 1996, Art. 16, § 4, the workload is made up of actions to prepare, monitor and evaluate pedagogical practice that involves, among other things, the preparation of plans for curricular activities, tests and other schoolwork [1]. Despite being provided for by law, the time needed to carry out these activities effectively can require exhausting working hours, even at home. However, digital technologies are allies in providing resources that can be used in the educational context, for example, to improve assessments [2].

A common practice is to search the internet for support materials to compose their lessons and teaching tools [3]. However, in 2019, the ICT Education survey strongly suggested that the pressure or lack of time to fulfill the content foreseen in the teaching-learning process was a barrier perceived by 79% of public-school teachers [4]. Santos and Sobrinho [5] have already pointed out that reducing teachers' working hours is essential for improving the quality of their work. In 2020, research from the same source highlighted that, with the distancing and greater insertion of digital technologies in the educational context, especially during the Covid-19 pandemic, there was an increase in teachers' workload, perceived by 73% of institutions. In addition, 61% of school managers cite a lack of ability on the part of teachers to use technological resources in didactic-pedagogical activities [6].

In this context, it is believed that teachers' workload and reliance on technical requirements and specific skills can be reduced and supported through tools that simplify the mnemonic processes associated with the use of digital technologies in education. The absence of a systemic solution that addresses this need constitutes the primary motivational impetus for the development of this research. Therefore, this study aims to design and implement a system named "OqKay", intended to automate searching and extracting questions from college entrance and official exams available in online repositories. The guiding inquiries of this investigation are: i) "What strategies facilitate the extraction of questions from college entrance exams available on the internet?" and ii) "What is the teachers' perception regarding the reduction of effort provided by a systemic approach in question extraction?".

The development of this study is presented in four additional sections: Sect. 2 highlights the work related to automated data extraction from internet resources; Sect. 3 presents the automated development and extraction techniques and the validation procedures of the systems approach; Sect. 4 discusses the results and discussions; and finally, Sect. 5 offers the concluding remarks, limitations, and implications of the system's approach for selecting and extracting questions from online educational exams.

2 Related Works

"OqKay" reflects a system approach aimed at automating the selection of questions from admission exams and official tests available online in Portable Document Format (PDF). Considering that most exams are available in PDF format, for a feasible and current solution approach, we analyzed studies that aimed to solve similar problems, namely the extraction of questions from online exams in PDF format.

In their research, Deon [7] developed a tool to automate the extraction of exam questions from PDF documents, using the Brazilian Informatics Olympiad exams as a database. The study explored different methods of extracting data from PDFs, employing libraries from the Python language. The pdfoftext library proved efficient in text extraction but was ineffective for content in figures and tables. The author turned to the LAREX computer vision software, which performs document segmentation into distinct regions, such as text, images, and tables, based on ground truth information to overcome this limitation. Ground truth is a geoprocessing term indicating areas manually demarcated or with computer vision tools. [8]. Creating a ground truth is a complex process, and despite software facilitation, specialized human intervention is necessary for adjustments. In the study by Wiechork (2021), a partially automated method is proposed based on the National Student Performance Exam (Enade) assessments and document segmentation. This segmentation was carried out using the Aletheia software, which, due to the absence of a command-line interface or programming frameworks, required specialized manual intervention for installation, configuration, and operation.

Although the examined solutions have shown promising results, both present the need for intervention by a specialized professional. The use of external software, employing computer vision techniques for document segmentation, implies the need for proper installation by users. Such a requirement may result in excluding individuals who need more technical knowledge, as may be the case for some teachers [3]. This dependency and the inherent limitations of existing solutions motivated the development of a project aimed at a semi-automatic system approach. This approach seeks to minimize the effort and time spent by teachers in extracting vestibular exam questions from online repositories, eliminating the need for specialized technical knowledge to perform this task.

3   Method

To develop a system tailored for educators, which obviates the need for technical or specialized knowledge, a comprehensive review of the state-of-the-art and relevant techniques was conducted. This review laid the foundation for developing the "OqKay" system. As a result of the analyses carried out during the review, the Python programming language was chosen to develop the "OqKay" system. The preference for Python is attributed to our familiarity with its daily use and the availability of libraries such as Scrapy, which provides various methods for Web Scraping suitable for various purposes. Regarding libraries for extracting data from PDF documents, PyMuPDF and pdftotext were examined. A detailed comparison between these tools is presented in Table 1.

Table 1. Python libraries for extracting data from PDF documents.

Features	PyMuPDF	pdfotext
Text extraction	Yes	Yes
Figure extraction	Yes	Not
Depends on external modules	Not	Yes

In the context of the present study, the PyMuPDF library is more comprehensive, incorporating modules capable of extracting both text and images from PDF documents, without additional modules or external software. As evidenced in the research conducted by Deon [7], the pdftotext library exhibited superiority exclusively in text extraction, disregarding images. However, for this investigation, the extraction of images proves crucial, as they can be employed in formulating questions.

The materials used as data sources for the "OqKay" system consisted of exams and answer keys from the Exame Nacional do Ensino Médio (Enem), which incorporate elements of verbal, non-verbal, and mixed language (text and images) in their items. The

Enem is an annual assessment conducted in Brazil, aimed at students who are completing high school, and its scores are used for access to higher education. The data extraction process from the Enem items was meticulously followed, as illustrated in Fig. 1.

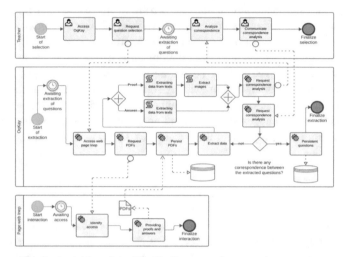

Fig. 1. Activities of the "OqKay" process for extracting questions.

The extraction of questions begins when the teacher accesses the OqKay system and requests the selection of questions. The system then accesses the website of the National Institute of Educational Studies and Research Anísio Teixeira (Inep) and automatically, using Web Scraping techniques, collects the files in PDFs. The website provides all the tests and respective answer sheets from previous editions of Enem. The files are persisted in a database according to test application editions and are then analyzed using the Ghostscript library.

3.1 Proposed Process

Web Scraping: This initial phase of the process involves the automated collection of exam papers and answer keys from the National High School Exam (ENEM) via the National Institute for Educational Studies and Research Anísio Teixeira (INEP) portal. The portal provides access to all editions of the ENEM exams and their respective answer keys.

File Storage: The collected documents, including exams and answer keys, are stored locally and organized in folders categorized by the year the exam was conducted.

Exam Processing: (1) Textual Segmentation: This begins with extracting textual content from the PDF files of the exams. A regular expression is used to segment the text, identifying the statements and options of the questions, as well as information about the edition, booklet, and corresponding page of each question. These data are later correlated with images during the image-matching phase. (2) Image Extraction: The images contained in the PDF files are extracted and stored in specific folders for each document. Information such as the edition, booklet, and page of the image are also stored for the subsequent phase of image matching.

Answer Key Analysis: The procedure for extracting text from the PDF answer keys follows the same methodology applied in the exam analysis. A regular expression is employed to capture and segment the answer key data.

Image Matching: In this stage, an interface is provided to the teacher to associate the images extracted in the previous phase with the questions identified in the textual segmentation of the exams. The user must determine to which question each image belongs. Each image is preliminarily linked to the questions from the same exam and page to facilitate this task. Considering that each page contains up to three questions, this is the maximum number of options per question for the user. The system also allows the user to associate multiple images with a single question, specifying whether they belong to the statement or the options. If there is no correspondence, the process returns to data extraction. Upon confirming the image matches, the system retains the questions extracted from the tests in a database and concludes the extraction process.

3.2 "OqKay" Assessment

Following the development of a solution to automate the Enem exam question search and selection, we validated this approach with educators. This stage involved collecting their insights and the advantages in time and effort optimization for searching, reading, analyzing, identifying, and extracting test questions from the web. Our hypotheses were: H0: a systemic data extraction method from PDFs enhances teachers' efficiency in selecting online test questions; Ha: a systematic approach to PDF data extraction has no impact on teachers' selection of online test questions.

The "OqKay" system was evaluated with 18 participants possessing higher education in diverse fields. Each participant engaged in a usage cycle of the app to analyze 20 test questions randomly chosen by the system. The "OqKay" system's graphical user interfaces (Figs. 2, 3, 4, and 5) displayed the PDF document for extraction (Fig. 3), enabling users to compare the extracted data with the original content. However, only printed Enem exams from 2011–2013 and 2016–2021 were considered, encompassing nine editions and 18 exams (two per edition) with 180 questions each, totaling 1,620 questions. Exams from 2009, 2011, 2014, and 2015 were excluded due to reading issues, and tests prior to 2009 were omitted because of layout variations.

Upon completing the selection and review of results, each participant filled out the System Usability Scale (SUS) questionnaire, evaluating effectiveness (achievement of user goals), efficiency (required effort and resources), and satisfaction (user contentment) [9].

4   Results and Discussions

The following sections present the main results regarding the system's strategies for extracting questions from the Enem exams. Moreover, teachers' perception of the "OqKay" system concerning usability and optimization of time and effort for question selection.

4.1 Strategies that Help Extract Issues

We developed strategies to assist teachers in selecting exam questions from the Brazilian National High School Exam (ENEM) available on the internet, using an approach from the "OqKay" system. To assess the algorithm's functionality, we provided an application on a web server. Its availability as a web application allows access through any web browser and does not require additional software installation on users' personal computers for data extraction. A set of functionalities was made available through the application interfaces, which will be described subsequently. The functionality to search for ENEM editions (Fig. 2) enables data extraction from the selected year's edition.

Fig. 2. Search screen for an edition of Enem.

Clicking the "Search" button, the system extracts files related to the selected edition. After the extraction, teachers are directed to a graphical interface displaying the extracted questions and a view of the original collected PDF document. This allows for reviewing the questions about the original document and verifying the accuracy of the extraction. The questions are formatted according to their attributes: shared text (if any), statements, alternatives, and images. This functionality corresponds to Web Scraping, file persistence, and test analysis phases. Figure 3 presents an example of question extraction performed by the system.

Fig. 3. View screen of extracted questions

Document segmentation driven by regular expressions enabled the search and validation of textual patterns, which, as Jargas [10] expressed, can be variable. In the case of ENEM exams, the perceived textual patterns were the number of questions, essay, and alternatives. The question number is always preceded by the word "question". The statement is a block of free text, all questions are multiple-choice, with five alternatives numbered from letter (a) to letter (e). With this knowledge, the strategies followed the enumeration of questions using a regular expression that will segment the text following the pattern "Question no". The free block that follows corresponds to the essay and question alternatives. Thus, they obtained their corresponding number and textual body for each test question. The alternative statements were also captured separately when refining the search. In certain cases, texts are used for multiple questions. In this scenario, a text segment applying to both questions is positioned immediately after the texts. Each shared text is introduced with the phrase "Text for the questions", including information specifying the questions it pertains to. Consequently, the method for establishing the rule for question identification is analogous to that employed for recognizing shared texts. It involves setting up a pattern of variables capable of detecting the presence of text related to the questions, concluding shortly before the start of the questions themselves.

Clicking the "Import" button in the Images section lets the user view images corresponding to the question. The teacher can associate the images extracted from the page with the question. The correspondence of the images can be analyzed at this stage, as shown in Fig. 4, during the extraction. From these strategies, we observed that some questions feature figures in the statement's body or the alternatives. This approach required a supervised strategy validated by the teacher, as it was impossible to automatically relate the extracted images to the questions they belonged to.

Fig. 4. Preview screen and import images corresponding to the question

Similarly, for the question screens, teachers can view the extracted templates for each question and thus compare them with the original documents. Figure 5 shows the screenshot related to viewing and an example of templates extracted by the "OqKay" system.

For this strategy, it was noted that the test answer sheets also follow a specific pattern, arranged in tables that relate the question numbers to the letter of the correct alternative. The digits represent the question number, and the answer sheets are represented by a letter belonging to the set [A, B, C, D, E] or by the word "canceled", representing a canceled question. In the model documents, the repetition of segments formed by pairs of text: the first from the category of digits and the second from the category of letters, enabled the creation of a regular expression to recognize the repetitions and capture the framed values. Table 2 presents the regular expressions for the strategies.

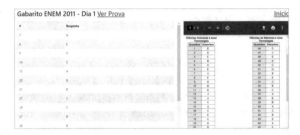

Fig. 5. Display screen for extracted templates

Table 2. Regular expressions are used in segmentation.

#	Regular expressions
1	#regex - extracts day and corresponding notebook from documents
2	dia_caderno_regex = r'(\d)(?:[°|°] dia) (?:|)*Caderno (\d{1,2})'
3	#regex - identifies questions in the text and groups them
4	corpo_questao_regex =
5	r'(Questão\s{1}\d{1,3})([\s\S]*?)(?=Questão\s{1}\d{1,3})|(Questão\s
6	{1}\d{1,3})([\s\S]*)(?:*.*?*)*)|(Questão\s{1}\d{1,3})([\s\S]*)'
7	#regex - identifies statement and alternatives for each question
8	enunciado_alternativas_regex = r'(.*) (A) (.*)(B) (.*)(C) (.*)(D)
9	(.*)(E) (.*)|(.*)'
10	#regex - identifies the question number
11	numero_questao_regex = r'(?:Questão\s{1})(\d{1,3})'
12	#regex - identifies whether there is shared text(s) and captures
13	them
14	texto_compartilhado_regex = r'para as questões (\d{1,3}) e
15	(\d{1,3})([\s\S]+?)(?=Questão)'
16	#regex - identifies the day of the test to which the answer sheet
17	applies
18	dia_gabarito_regex = r'(\d){1}[°|°] dia'
19	#regex - identifies the notebook to which the template applies
20	caderno_gabarito_regex = r'Caderno (\d{1,2})'
21	#regex - identifies the answer keys
22	resposta_gabarito_regex =
23	r'(\d{1,3})[\s](A|B|C|D|E|anulada|anulado)(?![\S])[\s]*(A|B|C|D|E|
24	anulada|anulado)?'

4.2   Teachers' Perceptions When Using the "OqKay" System to Extract Questions

To collect the teachers' perceptions, tests were conducted on 18 participants. By accessing the "OqKay" system through the web browser, users searched for specific evidence. After uploading the results, they viewed the questions and answer sheets and counted the number of incorrect questions and answer sheets. Participants responded to the SUS questionnaires and reported the number of errors immediately after each extraction interaction. A score was calculated based on the participants' responses, and we obtained an average of 87.08 in the scores assigned. Considering the total score given by participants to be above the average of 68 points, this represents good acceptance by participants of the solution developed in this study. About the assumption of the need for specific knowledge to select questions, the developed solution offers an abstraction of repetitive steps to facilitate teachers' activities.

The nine editions of Enem considered in the study total 1,620 questions. Each of the 18 participants analyzed 20 extracted questions; this represents 360 (22%) revised questions. The results show the count of errors reported by participants considering: (i) empty, incomplete or different statements from the original document; (ii) empty, incomplete or different alternatives to the original document; (iii) question that has shared text: shared text that is empty, incomplete, or different from the original document; (iv) question that contains images, on the screen to view and import images corresponding to the question does not display any image; (v) empty template or different from the reference document. In this sense, the participants found errors related to the statement, alternatives, shared texts, and images. The accounting of errors is found in Table 3.

Table 3. Results of the extraction of test questions

Editions (test days)	Questions	Evaluated	Hits	Errors
On the first day of the test	810	180	163 (90.6%)	17 (9,4%)
2nd day of test	810	180	85 (47,2%)	95 (52,8%)
Total	**1620**	**360**	**248 (68,9%)**	**112 (31,1%)**

Among the 360 questions evaluated, 68.9% were correct. For the tests on the first day, a significantly higher percentage of correct answers (90.6%) was observed concerning errors (9.4%). For the tests on the second day, there is a lower percentage of correct answers (47.2%) than errors (52.8%). We noticed that this occurs because the tests on the second day have more figures, graphic representations, or formulas (due to the organization of the test editions containing questions from the areas of Exact Sciences and Nature). In these cases, the system approach cannot identify the formulas as text or images.

As the documents relating to the test templates were made available separately, the participants viewed the data extracted from the templates on a separate screen, and the items were counted. The answer rate for the answer sheets was 100%. Some factors contributed to the excellent result: the consistency of the strategies used and the effectiveness in collecting the layout of the templates; the absence of images and figures; the amount of content present in these documents is smaller than the test documents.

5   Final Considerations

The "OqKay" system approach benefits educators by reducing the time and effort spent searching, reading, analyzing, identifying, and extracting test questions from the Internet. We facilitate the distribution of the developed solution to automate and optimize teachers' routine time searching for questions to compose their teaching-learning artifacts. As a result, the system was well-received by the educators who tested it, achieving a performance rate of 68.9% in extracting correct questions.

The limitations of this study include the extraction of figures, graphical representations, or formulas in questions from the fields of Exact and Natural Sciences, as well as the connection between regular expressions and the layout of the standard test document. It is anticipated that subsequent research may continue to explore current techniques for extracting data from PDF documents to develop a more comprehensive solution that can encompass data categories not covered in this study. Furthermore, applying regular expressions in segmenting textual data from PDF documents warrants more in-depth analysis. This includes extending the research to other databases and exploring methods for parameterizing acceptable patterns, aiming to enhance the efficacy and applicability of the method.

Acknowledgements. This work was supported by the FCT – Fundação para a Ciência e a Tecnologia, I.P. [Project UIDB/05105/2020].

References

1. Pernambuco. Lei nº 11.329, de 16 de janeiro de 1996, Estatuto do Magistério Público de Pré-Escolar, Ensino Fundamental e Ensino Médio do Estado de Pernambuco. Assembleia Legislativa do Estado de Pernambuco (1996). https://legis.alepe.pe.gov.br/texto.aspx?tipono rma=1&numero=11329&complemento=0&ano=1996&tipo=&url. Accessed 16 May 2022
2. Bessa, A., Silva, D.R.C.: Multiprova: aprimorando a avaliação com o uso da tecnologia. RENOTE **15**(1), 1–10 (2017)
3. da Silva, C.J.P.: Design de um sistema de informação para apoiar a atividade de planejamento de aulas: uma abordagem situada. Ciência da Computação, Universidade Federal de Pernambuco, Recife, PE, Brasil, Dissertação de Mestrado (2020)
4. Centro Regional de Estudos para o Desenvolvimento da Sociedade da Informação (CETIC). TIC Educação 2019. CETIC; NIC; CGI, 21 de novembro de (2020). https://cetic.br/pt/publicacao/pesquisa-sobre-o-uso-das-tecnologias-de-informacao-e-comunicacao-nas-escolas-brasileiras-tic-educacao-2019/. Accessed 16 May 2022
5. dos Santos, A.A., Sobrinho, C.L.N.: Revisão sistemática da prevalência da Síndrome de Burnout em professores do ensino fundamental e médio. Revista baiana de saúde pública, vol. 35, no. 2, p. 299 (2011). https://doi.org/10.22278/2318-2660.2011.v35.n2.a307
6. Centro Regional de Estudos para o Desenvolvimento da Sociedade da Informação (CETIC). TIC Educação 2020 - edição COVID-19, metodologia adaptada. CETIC; NIC; CGI, 25 de novembro de (2021). https://cetic.br/pt/publicacao/pesquisa-sobre-o-uso-das-tecnologias-de-informacao-e-comunicacao-nas-escolas-brasileiras-tic-educacao-2020/. Accessed 16 May 2022
7. Deon, O.O.: Automatizando a exportação de questões de provas da olimpíada brasileira de informática por meio de ferramentas de extração de texto e visão computacional. Trabalho de Conclusão de Curso, Bacharel em Ciência Computação, Universidade Federal de Santa Maria, Santa Maria, RS, Brasil (2018)
8. Wiechork, K.: "Extração automatizada de dados de documentos em formato PDF: aplicação a grandes conjuntos de exames educacionais", Dissertação de Mestrado. Universidade Federal de Santa Maria, Santa Maria, RS, Brasil, Dept. Ciência da Computação (2021)
9. Brooke, J.: SUS: a retrospective. J. Usability Stud. **8**(2), 29–40 (2013). https://doi.org/10.5555/2817912.2817913, Bloomingdale, IL
10. Jargas, A.M.: Expressões Regulares: uma abordagem divertida. 4ª ed. rev. e ampl. São Paulo, SP: Novatec Editora, p. 220 (2012). ISBN 978–85–7522–337–6

Building and Analyzing an Open Educational Resource Repository on the Inovaula Platform

Rafael L. S. M. Albuquerque[1], Alex Sandro Gomes[1], Leandro Marques Queiros[1], Carlos José Pereira da Silva[1], and Fernando Moreira[2]($\boxtimes$)

[1] Centro de Informática, Universidade Federal de Pernambuco, Recife, Brazil
{rlsma,klss,asg,lmq,cjps}@cin.ufpe.br

[2] REMIT, Universidade Portucalense, Porto and IEETA, Universidade de Aveiro, Aveiro, Portugal
fmoreira@upt.pt

Abstract. Given the increasing adoption of Digital Information and Communication Technologies (DICTs) in teaching practices, it is observed that there are still practices that do not fully benefit from the culture of sharing and reusing Digital Educational Resources (DERs) available on the internet. In this study, a new repository integrated into a lesson planning support platform was developed and evaluated. To assess the efficacy of the repository, a semi-structured interview, and the System Usability Scale (SUS) questionnaire were used, aiming to measure usability. The results indicated satisfactory usability, with an average score of 86.96. When asked about the utility of the repository, participants perceived that the system positively contributes to the reuse of educational resources in the classroom.

Keywords: open educational resources · open educational resource repository · usability

1 Introduction

The COVID-19 pandemic accelerated the adoption of Digital Information and Communication Technologies (DICTs) in teaching practices. On the other hand, data collected by CETIC.BR [1] showed a deficit in teacher training for using DICTs. The research identified that 61% of schools participating in the study reported needing teacher skills in using technology resources. This deficit can hinder teaching-learning practices that benefit from access to Open Educational Resources (OER) available on the Internet.

Given this scenario, the Open Education (OE) movement is growing, encouraging collaborative and open practices to benefit quality education. The movement creates a culture based on the manipulation, reuse, and sharing of OER to make teaching configurations more flexible, create access opportunities, and democratize learning for all people. Taking OER is the initial step toward a culture of sharing and transparency [2].

As per the definition provided by the United Nations Educational, Scientific and Cultural Organization (UNESCO), the concept of OER categorizes any educational

Á. Rocha et al. (Eds.): WorldCIST 2024, LNNS 988, pp. 43–52, 2024.
https://doi.org/10.1007/978-3-031-60224-5_5

material intended for use in teaching and learning. It is openly accessible for utilization and modification by external individuals, without the requirement to pay royalties or obtain licenses [3].

The use of OER is an alternative to overcoming the culture of passivity in educational practices. The objective is to encourage teachers and students to interact with each other in the production and adaptation of educational resources, inserting them into their context [4]. Evidence shows that using these resources allows positive results for the performance of Basic Education students [5].

Although OERs are essential tools to enhance teaching-learning, teachers still need help locating them on the Web. Traditional search engines, for example, Google Search, present, in response to searches, a large volume of irrelevant content protected by copyright laws, making it difficult for students or teachers to access educational content [6].

In this context, Open Educational Resources Repositories (OERR) are a viable solution to simplify educators' access to resources and knowledge. The current scenario of OERR in Brazil was raised by Medeiros et al. [7]. The author identified only two high-quality repositories, and only one allows the import of Digital Educational Resource (DER) for lesson planning, a condition that facilitates teaching practices.

Given the lack of possibility of importing resources directly into lesson planning, we formulated the following hypothesis: the use of an OERR that allows importing resources to a collaborative lesson plan platform is well accepted by Basic Education teachers. The present study aimed to create and analyze the effectiveness of an OERR and, to this end, evaluated the search and selection of resources in planning classes mediated by the Inovaula.com platform.

This document is organized as follows: In Sect. 2, the concept of OER reusability was presented; In Sect. 3, the method for creating the OERR, creating the REA database, and validating the proposed solution was described; while in Sect. 4, the results obtained with the new repository were presented; Finally, in Sect. 5, final considerations and perspectives for future work were presented.

2   Reuse of OER in Brazilian Education

For the effective reuse of OER, accurate localization is essential. To facilitate this, metadata is assigned to each OER with dual objectives: firstly, to categorize resources for easier future retrieval via the Internet, and secondly, to describe their functional and educational characteristics. Through such retrieval, these resources can be utilized to create diverse and enriching learning experiences or be integrated into various applications [8].

The categorization and retrieval of OER are conducted within OERR. Despite numerous Brazilian repositories, their utilization among educators remains suboptimal [9]. This is evidenced by teachers' infrequent or rare use of these resources [10].

The limited engagement of teachers with OERR may indicate a gap in integrating these repositories with other educational applications. For instance, a more seamless connection with Learning Management Systems (LMS) or tools that aid in lesson planning and preparation could enhance usability. As Fabre et al. [8] suggest, the efficiency

of resource utilization is significantly improved when resources are systematically organized and classified in repositories that are capable of integrating with various computer
systems.

3 Method

The realization of the repository in question was carried out based on a set of requirements
outlined in Silva's [11] study. This study is grounded in preceding research that delineated
the design of a repository for OERs integrated into a lesson planning support system.
The researcher conducted an ethnographic analysis of lesson planning practices and
developed a prototype application to assist in this activity, based on the needs identified
in teaching practices.

Building on this conceptual prototype, a preliminary low-fidelity version was evolved
into an enhanced high-fidelity version, compatible with web browsers. Subsequently,
usability tests were conducted for the validation and refinement of the prototype.
Concluding this phase, the implementation of the OER repository was carried out in
accordance with the established specifications.

To evaluate the implemented OERR, a collection of OERs from an existing resource
repository was gathered, aiming to constitute the database of the repository. Usability
tests of the reuse system were conducted with a representative group of Basic Education
teachers and Information Technology professionals, aiming to assess the efficacy and
effectiveness of the proposed solution.

3.1 Repository Implementation

To implement OERR, the JavaScript language, React.js libraries were used to develop
the graphical interface, and Node.js with Express and the PostgreSQL database to persist
and retrieve application data. The front end (visual part of the website) was built using the
Single-Page Application (SPA) concept, which provides a usability experience without
updating entire pages, loading only part of the content, just like smartphone applications
do. In contrast, the back-end (data persistence part) used a Model View Control (MVC)
architecture to facilitate the code's organization and reading. Furthermore, all processing
was delegated to the relational database using the Sequelize tool.

The implementation of the main screens and their respective visual components was
developed based on the high-fidelity prototype, adhering to the following Functional
Requirements (FR): FR1 - to list OER available on the platform; FR2 - to search for
OER by title, description, or keywords; FR3 - to filter OER by competencies of the
National Curricular Common Base (BNCC), resource type, or evaluation rating; FR4
- to view OER, including metadata visualization, download or access to the resource,
and a list of related OER; FR5 - to evaluate the educational resource with a rating of
0–5 and provide a rationale; FR6 - to report a problem with a selected OER; FR7 - to
link an educational resource to a user's lesson plan; RF8 - to publish a new educational
resource.

3.2   Resource Base Creation

The data extraction, transformation, and loading processes were used to extract REA from another repository to create the solution's database. The technologies selected for this part of the project were Python 3 and Jupyter Notebook with the NumPy and Pandas libraries for tools that facilitate data extraction, analysis, filtering, and transformation.

The initial stage was the selection of the OERR, from which the resources were collected. Then, the MEC Integrated Platform was chosen, as it was the OERR best evaluated by Medeiros et al. [7] and had an open web service. Then, thousands of requests were made to the repository's API, collecting over 300 thousand resources and saving them on the computer's hard drive running the script.

In the transformation stage, the saved resources most relevant to the research context were filtered: OER in Portuguese, OER in English intended for learning English, and OER with Creative Commons licenses. After filtering, the attributes of the Integrated Platform's REA data model were mapped to the implementation data model of this study. The code generated by this step is in the public domain on the Internet through GitHub.

The resources were converted to JSON format after mapping to the implementation data model. A JavaScript script was created using the Sequelize seeders concept, loading the data directly into the database.

3.3   OERR Usability Evaluation

This procedure aims to understand efficiency, effectiveness, and user satisfaction, and the metric that covers these parameters in a system is usability, according to ISO 9241–11 [12]. To quantify it, the participant was inserted into a context without the need to configure an account and lesson plan to test only the functionalities implemented by the study.

A set of activities to be carried out in the system was presented, and after the activities, each participant responded to a SUS—System Usability Scale [12] questionnaire.

The SUS is widely used to evaluate the perception of a system's usability. It consists of 10 standardized items alternating between positive or negative tones, with five response options [13]. A questionnaire was constructed using the Likert scale—which varies between 1, meaning I completely disagree, and 5, meaning I completely agree—and generates a score from 0 to 100. The procedure defined was:

- Find the "A Walk in the Museum" resource and link it to a "History of Museums" lesson plan;
- Rate the resource "A Walk through the Museum" with five stars;
- Report the "URL unavailable" problem in the "A tour at the Museum" resource;
- View any video-type resource;
- View a resource with five stars;
- Publish a new resource by filling in the fields as follows:

  - Feature Title: Usability Testing
  - Feature Overview: Usability Testing
  - Author of this resource: Test
  - Other authors: Leave blank

- Resource Type: Image
- Education level: Elementary school
- Area of knowledge: Art
- License: CC BY-SA
- Resource file

The result was calculated by adding the individual contribution scores of the items and multiplying by 2.5. For even items, this contribution was calculated by subtracting the user's response from 5 and, for odd items, subtracting it from the response from 1 [12].

The standard questionnaire items were constructed in English, so to maintain the validity of the SUS for Portuguese, the questions translated and cross-culturally adapted from [14]: 1. I would like to use this system often; 2. I found this system unnecessarily complex. 3. I found this system easy to use. 4. I thought I would need help from a technical person to be able to use this system. 5. I thought the various functions of this system were well integrated. 6. The system has much inconsistency. 7. Most people can learn to use this system quickly. 8. I found this system very cumbersome to use. 9. I felt very safe using the system. 10. I needed to learn many things before I could use this system.

In addition to the SUS questionnaire, the final evaluation form contains questions to identify the user's demographic profile, area of activity, degree of familiarity with technology, and degree of familiarity with OER.

Twenty-three voluntary participants were selected, fulfilling the minimum sample required to validate the result [15]. Among the participants, we had 11 IT professionals (Programmers and UX Designers), 10 primary school teachers, and two students. All were randomly recruited via social media.

Finally, a semi-structured interview was conducted with the participants after completing the form. The objective was to identify the most significant difficulties in using the solution and the perception of whether linking OER to lesson plans would increase the reuse of digital resources in the classroom. It is important to note that for participants with little or no experience in lesson planning, such as IT professionals, a general context was given for creating and using lesson plans.

4 Results

The following sections present the main results obtained in developing the OERR. In addition, the evaluation results carried out with IT professionals, teachers and students about the resolution's usability and ease of use developed.

4.1 Concept Implementation

The repository was implemented to meet all requirements listed in Sect. 3.1. The requirements were distributed across three screens: listing, viewing, and publishing OER. Figure 1 displays the repository's main screen, and it is from there that it is possible to navigate between the other screens. On this screen, you can view a list of all

OERs available on the platform (FR1); perform a textual search by title, description, or keywords of the OER (FR2); filter OER by BNCC competencies, type of resource or evaluation score (FR3); and linking a resource to a user lesson plan (FR7).

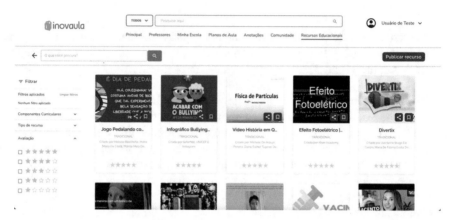

Fig. 1. Repository main screen

Figure 2 shows the screen for publishing a new resource, accessed through the "publish resource" button on the main screen. On this screen, it is possible to publish a new OER in two steps (FR8): in the first step, the resource metadata is filled in, and the resource file is uploaded; in the second, the most appropriate license is chosen from the available options.

Fig. 2. New REA publication screen

The OER visualization screen, Fig. 3, presents all information about the accessed resource, enabling access or download (FR4), resource evaluation (FR5), reporting content-related errors (FR6), and linking the resource to user lesson plans (FR7). A list of related resources is also shown, constructed from common keywords.

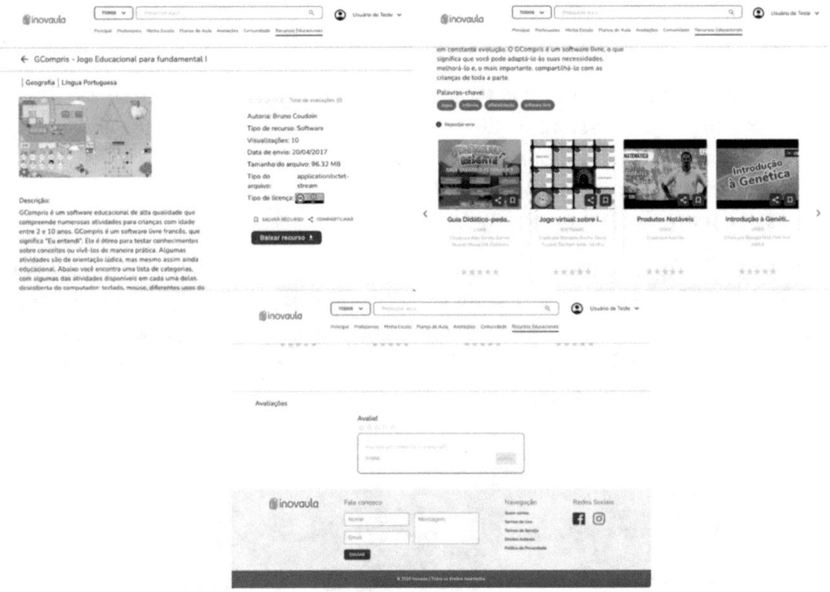

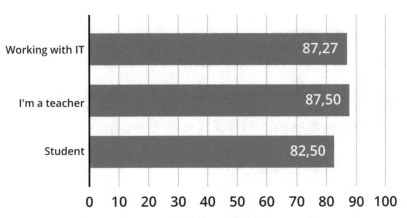

Fig. 3. Resource preview screen

4.2 Usability Assessment

The average score obtained was 86.96, a result classified as excellent [16]. Excellent results were obtained by profile, with no significant difference between the two main study groups (Fig. 4).

Fig. 4. Bar graph with average SUS score by profile

Boucinha & Tarouco [17] propose a relationship between the SUS questions and the components that define usability according to Nielsen [18]:

- The system's ease of learning is represented in items 3, 4, 7, and 10 of the SUS;
- The efficiency of the system is represented in items 5, 6, and 8 of the SUS;
- Ease of memorization is represented by item 2 of the SUS;
- Minimizing errors is represented by item 6 of the SUS;
- User satisfaction is represented by SUS items 1, 4, and 9.

The graph in Fig. 5 shows the SUS score by profile and usability quality component. IT professionals believe that the system is more efficient and has fewer inconsistencies. This result may be because this group is more familiar with Web interfaces. On the other hand, teachers believe that the system is easy to memorize, as they are more familiar with the concepts and terms found in the system.

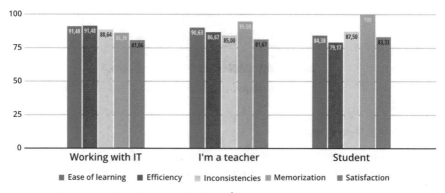

Fig. 5. Graph with average SUS score by quality component and profile

During the interviews and usability tests, improvements in the solution's usability were identified: cleaning fields after submitting the action, creating a tour or manual, improving the positioning of OER information in the visualization, and adding a preview of the resource.

Finally, all participants realized that linking an OER to a lesson plan would increase the reuse of digital resources in the classroom. Furthermore, 65% of them mentioned that the solution would increase teachers' productivity in preparing classes.

5   Conclusions and Future Work

The present study aimed to develop and evaluate the effectiveness of an Open Educational Resources Repository (OERR) within a lesson planning platform, focusing on user experience and usability perception. Additionally, to enhance productivity and facilitate the reuse of digital resources in educational environments, the repository was integrated into the Inovaula.com platform. Based on prior specifications, the system was implemented, and its efficacy was assessed through usability tests and interviews.

The research revealed that the OERR possesses remarkable usability, achieving an average score of 86.96 on the System Usability Scale (SUS) questionnaire. All participants acknowledged that the functionality of linking Open Educational Resources (OER) to lesson plans enhances the reuse of these resources in the classroom. This finding aligns with the study by Fabre et al. [8], highlighting the efficiency of repositories that can be integrated with other computer systems.

It is proposed to correct and reevaluate the usability issues identified in the interviews and tests for future directions. Furthermore, to enrich and diversify the collection of OER in the solution, it is considered essential to extract educational resources from other OERRs and integrate them with tools for creating and editing OER.

References

1. Centro Regional de Estudos para o Desenvolvimento da Sociedade da Informação (CETIC), TIC Educação (2020). https://www.cetic.br/media/analises/tic_educacao_2020_coletiva_i mprensa.pdf. Accessed 05 Jul 2022
2. Amiel, T.: Educação aberta: configurando ambientes, práticas e recursos educacionais. em Recursos educacionais abertos: práticas colaborativas, políticas públicas, Salvador: Edufba, pp. 17–33 (2012)
3. Butcher, N., Kanwar, A., Uvalic-Trumbic, S.: A basic guide to open educational resources (OER). Commonwealth of Learning, UNESCO (2011)
4. Starobinas, L.: REA na educação básica: a colaboração como estratégia de enriquecimento dos processos de ensino-aprendizagem. em Recursos educacionais abertos: práticas colaborativas políticas públicas, Salvador: Edufba (2012)
5. de Santos, E., Oliveira, F.K., Gomes, A.S., de Brito, J.J.O.R.T.: Mapeamento sistemático acerca das práticas docentes com o uso de Recursos Educacionais Abertos. In: Proc. Brazilian Symposium on Computers in Education (Simpósio Brasileiro de Informática na Educação-SBIE), no. 263 (2017)
6. de Souza, R.C., Neto, F.M.M.: Construção de um Repositório de Recursos Educacionais Abertos Baseado em Serviços Web para Apoiar Ambientes Virtuais de Aprendizagem. RENOTE, vol. 12, no. 2 (2014)
7. Medeiros, R., Duarte, M., Viterbo, J., Maciel, C., Boscarioli, C.: Uma Análise Comparativa entre Repositórios de Recursos Educacionais Abertos para a Educação Básica. In: Anais do XXXII Simpósio Brasileiro de Informática na Educação, SBC, pp. 213–224 (2021)
8. Fabre, M.-C.J.M., Tamusiunas, F., Tarouco, L.M.R.: Reusabilidade de objetos educacionais. RENOTE, vol. 1, no. 1 (2003)
9. Oliveira, F.K., Abreu, K.F. da Silva Gomes, A.A.: Formação profissional em recursos educacionais abertos. Revista Semiárido De Visu, vol. 3, no. 2, pp. 98–109 (2016)
10. Santana, S.C., Lopes, C.V.: Objetos de Aprendizagem: Visão do docente na efetividade do processo de ensino-aprendizagem na educação superior (2019)
11. da Silva, C.J.P.: Design de um sistema de informação para apoiar a atividade de planejamento de aulas: uma abordagem situada. Ciência da Computação, Universidade Federal de Pernambuco, Recife, PE, Brasil, Dissertação de Mestrado (2020)
12. Brooke, J.: SUS - A quick and dirty usability scale. In: Usability Evaluation in Industry, pp. 189–194 (1996)
13. Lewis, J.R.: The system usability scale: past, present, and future. Int. J. Human-Comput. Interact. **34**(7), 577–590 (2018)
14. Lourenço, D.F., Carmona, E.V., de Moraes Lopes, M.H.B.: Tradução e adaptação transcultural da System Usability Scale para o português do Brasil. Aquichan, vol. 22, no. 2, art. 4 (2022)

15. Stetson, J.N., Tullis, T.S.: A comparison of questionnaires for assessing website usability. apresentado na UPA (2004)
16. Bangor, A., Kortum, P., Miller, J.: Determining what individual SUS scores mean: Adding an adjective rating scale. J. Usability Stud. **4**(3), 114–123 (2009)
17. Boucinha, R.M., Tarouco, L.M.R.: Avaliação de ambiente virtual de aprendizagem com o uso do SUS-System Usability Scale. RENOTE, vol. 11, no. 3 (2013)
18. Nielsen, J.: Usability 101: Introduction to Usability (2003). https://www.nngroup.com/art icles/usability-101-introduction-to-usability/. Accessed 17 Aug 2022

L&D Model of Information Technology Application for Remote Teaching and Learning in Physical Education at Technical University

Maryna Litvinova[(✉)] [iD], Oleksandr Shtanko [iD], Iryna Smirnova [iD],
Svitlana Karpova [iD], and Viktor Prytula [iD]

Kherson Educational-Scientific Institute of Admiral Makarov National University of
Shipbuilding, Kherson, Ukraine
`{maryna.litvinova,oleksandr.shtanko,iryna.smirnova,`
`svitlana.karpova,viktor.prytula}@nuos.edu.ua`

Abstract. Currently, the organization of remote learning in physical education based on the latest possibilities of information technology in higher education institutions is heuristic. The purpose of the work is the development and approval of the model of the use of information technology for the implementation of the remote educational process in physical education in institutions of higher technical education. The "Learning and Development" (abbreviated "L&D") model, formed in business structures to ensure professional development, was taken as the basis of the model for the use of information technology. According to the L&D model, an action algorithm with four components was proposed. The first is the creation of a strategy for the use of information technology for effective remote learning in physical education. The second is to ensure the information competence of students and teachers for conducting online classes. The third is the creation of conditions for the individualization of remote learning. The fourth is to ensure reliable and objective monitoring of remote classes. The components and implementation conditions of each component were examined in detail and their approval was carried out. The L&D model of information technology use allows to create a system of information and technological support for the remote educational process in physical education and to increase the level of student satisfaction with the remote form of learning.

Keywords: Information support · Learning and Development · Remote teaching

1 Introduction

Pandemics, martial law and other reasons limiting the presence of students in educational institutions make the task of ensuring high-quality remote learning based on the latest possibilities of using information technology more urgent.

The overarching task of remote teaching in universities is not just to preserve the quality of education, but also to improve it compared to offline teaching. The most difficult task in terms of a number of indicators is the provision of high-quality remote

Á. Rocha et al. (Eds.): WorldCIST 2024, LNNS 988, pp. 53–63, 2024.
https://doi.org/10.1007/978-3-031-60224-5_6

teaching in physical education. It involves not only the transfer of teaching information, but also requires special conditions for online interaction between a teacher and a student, which provide an opportunity to safely and effectively implement information through practical actions. Therefore, the development of new educational models based on the latest possibilities of using information technology is relevant.

2 Background and Related Work

Information technology is considered as a set of hardware, means and methods that provide information management. According to researches [1, 2], this technology forms a fundamentally different style of teaching activity and sports training process, which are the most psychologically acceptable and comfortable for students. Competent application of information and computer technologies during PE classes ensures effective mobilization of creative opportunities, intellectual and physical potential of students. Many researches have been devoted to their implementation in physical, health and sports activities of educational institutions [3]. However, most of them are focused on offline classes or on a mixed form of classes: online and offline at the same time [4].

Remote teaching technologies in physical education can currently be divided into several main directions.

The first direction includes technologies using Internet platforms or Internet channels. Table 1 lists the names of some popular platforms that are mentioned in various researches [3, 5–8] and to some extent suitable for teaching physical education at the university. The table shows their functionality, terms of conducting classes and terms of payment for using the software. Most of software is applications that can be downloaded for free to a computer or mobile hardware. If the programs are paid, and the university does not have the opportunity to purchase a subscription for their use, this significantly limits the suitability of the corresponding platform for a given educational institution.

The YouTube platform (the last three lines in Table 1) contains a lot of free video channels or individual videos for online training. University teachers often post their videos on it. Some of them prefer their own Web-platforms [9] or post materials directly on the website of educational institution [10, 11].

The second direction is the introduction of special virtual reality programs in physical education and training. A survey of the use of such programs was carried out in work [12]. In research [13], the high efficiency of using virtual motion capture systems was shown. However, the use of programs for virtual training in the educational process is limited by the need for students to use their head-mounted displays (helmets). In addition, the most of such programs also requires additional sports equipment. Therefore, educational institutions currently use only the services of VireFit and Nike "Kinect Training" during remote teaching among many relevant programs. They are a virtual sports trainer, tracking and visualizing movements and providing recommendations. These programs have different game modes: basketball, tennis, boxing, etc. with different levels of complexity without requiring special equipment [14]. But the use of all types of the programs is paid.

A separate direction of the use of information technology is related to the control function of software for health-related physical education [15]. These are diagnostic, diagnostic-recommendation and management programs. According to the results of

Table 1. Popular Internet platforms and Internet channels for online PE classes.

Platform or channel	Programs Functionality	Conditions
Nike Training Club	Training for different training levels is designed to load different muscle groups. Any training can be adjusted to individual requirements	Indoor Free
Adidas Training	A large number of training sessions lasting from 9 to 45 min: classes with exercises for weight loss, muscle gain, improving various indicators	Indoor Free
Freeletics Training Coach	Short fitness training. Advertised as free, but the main options are paid (by subscription)	Indoor. Paid, subscription
Barry's Bootcamp	Fitness training	Indoor. Paid, subscription
IGContest	Gymnastics, platform diving, workout, snowboarding, yoga, pole dance, etc. After recording the performance of various elements, it is possible to get an assessment of such performance not only by the teacher, but also to participate in an online tournament	Indoor and outdoor Free
Workout Trainer	Various exercises for indoor and outdoor sports. It is possible to create a workout by yourself by adding individual exercises to "favorites"	Indoor and outdoor Free
Dartfish	Video analysis software, special programs when learning and correcting movements, has digital video graphics to use visual feedback without stopping training	Indoor. Paid, subscription
POPSUGAR Fitness	Short online workouts: cardio and functional training, dance aerobics, yoga, etc. (YouTube)	Indoor Free
Passion4Profession	Short online workouts for training all muscle groups (YouTube)	Indoor Free
The 12-min Athlete	Mobile application with high-intensity interval training (YouTube)	Indoor Free

special hardware, measuring and recording the remote traveled, walking pace, pressure, pulse rate and etc., the special software interacts with the user and provides information on the state of health based on the principle of feedback, gives personal tasks, monitors their safe execution.

The analysis of scientific works shows that at this time the organization of remote teaching in physical education in higher education institutions (HEIs) is heuristic. Each HEI has its own strategy for the use of information technology [16]. As a result, preference is given to one or another direction and the potential opportunities of others are lost. There is no general algorithm or model for the organization of the remote educational process. The development of a suitable model and its approbation during remote teaching is one of the key tasks of planning the educational process, both at the national and global levels, and is a way to improve physical education in general.

Hypothesis. The organization of remote teaching in physical education should be based on a theoretical and experimental educational model based on using information technology. Such a model should optimally use the technological capabilities of an educational institution of a certain type and ensure the highest possible level of remote teaching.

Purpose of the Study is the development and approval of the model of information technology application for the implementation of the remote educational process in physical education in institutions of higher technical education.

3 Method

The information basis using of computer technologies for building a physical education teaching model in technical universities was based on data obtained from publications in Scopus and the Web of Science Core Collection, as well as from the websites of leading technical universities. In addition, the materials of the leading web platforms containing video materials for online PE classes and offers of virtual reality programs for physical training were analyzed.

The research project was implemented at (*the Name of the Educational Institution will be given after the review*) during the fall and spring semesters of the 2022–2023 academic year. 167 first- and second-year students took part in the research.

The advantage of technical universities is the presence in the curricula of students of the computer science course, which is taught starting from the first academic semester and provides competence in using various types of software. For effective remote work, both the student and the teacher must be able to receive, process, systematize and use information using the necessary computer software and telecommunications tools.

An online survey of students was conducted to control remote learning in physical education, which took place according to the developed model. The following were used: a preliminary survey regarding the availability of online learning tools for students; a final survey to analyze the results of using the proposed learning model. To analyze the impact of the L&D model on the quality of remote learning, a correlation analysis using a multiple linear regression calculation model was used. The method of its application for various types of experiments was described in detail in a previous work [17]. Calculations were made in Excel.

4   Results

4.1   Model Description

The "Learning and Development" model (abbreviated as "L&D") was taken as the basis of the model for the use of information technology during remote learning in physical education, which was previously used in the KESI NUS for the information support of training in other disciplines [18, 19]. This model was transformed specifically for the conditions of remote learning.

The L&D model has historically been formed in business structures and defined the algorithm of actions for providing any professional development that the business provides to its employees. It has been used for career development, continuing studying at corporate universities, leadership development programs, skills training and talent management strategies.

Regarding the application of information technology for remote learning in physical education, the L&D model is considered as a way of organizing the teaching and learning process, which, according to a proven algorithm, creates conditions for improving PE classes, obtaining the necessary knowledge and skills to increase their effectiveness. This model can be implemented either centrally, through a common server of the university, or independently, within a certain group of students, or individually with each student separately.

The proposed algorithm of actions corresponding to the L&D model of information technology application during remote learning in physical education is shown in Fig. 1.

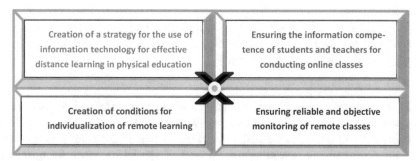

Fig. 1. Algorithm of actions for the information technology application during remote teaching and learning in physical education.

Let's consider in more detail all the components of the given algorithm of actions.

1. Creation of a strategy for the use of information technology for effective remote learning in physical education.

This strategy is implemented based on the available resources of the educational institution. An audit and analysis of the suitability for use of all means for conducting online PE classes available at the university is planned. The priority online platform, the necessary software for conducting classes on the basis of the university, the possibility of access to training programs through various Internet platforms, including virtual reality programs, are determined.

Based on the available resources, the optimal strategy for conducting online classes is determined and the corresponding training program is developed.

2. Ensuring the information competence of students and teachers for conducting online classes.

There is an integration of training programs in the disciplines "Physical Education" and "Informatics" with the aim of students obtaining the competence to use software, which is necessary for conducting remote PE classes. Through training, the appropriate competence of the teaching staff is also ensured. The main function of a teacher is determined as the function of a tutor, that is, an instructor and a personal mentor, which helps the student to better understand his/her capabilities and competently set priorities among the types of physical education activities proposed by the teacher and learn to study.

3. Creation of conditions for individualization of remote learning.

Students are considered as individual consumers of a certain educational service that promotes a healthy lifestyle. Conditions are created for choosing an individualized thematic block containing certain types of physical exercises and physical training. The choice is made on the basis of personal physical capabilities, target positions and conditions for conducting classes. With the help of a teacher-tutor, students find out which exercises from the proposed list meet their needs and preferences. This is achieved by execution of preliminary mailings of information, surveys and personal online consultations.

During classes, students are divided not into academic groups, but according to a selected block of exercises, that is, according to the possibility of performing a certain type of physical activity at a certain time.

4. Ensuring reliable and objective monitoring of remote classes.

Control over the remote PE classes consists of two parts: safety and normative. The safety part is realized by familiarizing students with methods of self-monitoring of health and safety rules during physical activities. Students acquire the competence of using diagnostic programs and analyzing the readings of special hardware for operational assessment of the state of health. The second part consists in the teacher obtaining objective data on the implementation of standards for annual evaluation.

4.2 Approbation of the Model

Let's consider the implementation of the L&D model in the KESI NUS for each component of the algorithm of actions.

1. According to the analysis of the software available in the KESI NUS, an electronic library with a structured information system was created, which made it possible to provide:

   – online support for students and teachers and quick updating of learning materials;

  – optimized information search for each type of physical exercise and load with the possibility of reviewing the search process, comparing and evaluating the received information, identifying gaps in it.

The electronic library made it possible to implement functional requirements according to the L&D model, namely: ease of online information search and quick access to it; the ability to work with large arrays of information that are updated daily; confidentiality and security of working with information.

Based on student surveys about the convenience of using various learning platforms, the Google Classroom platform was chosen to store learning information, and the Zoom platform was chosen to conduct online classes. Training programs for physical education have been compiled.

A survey was conducted regarding whether students have access to electronic learning tools, equipment needed for physical education and sports classes (PESC), special equipment needed to work with virtual reality programs. The results of the survey according to the percentage of statements received are shown in Table 2.

Table 2. Results of survey.

Statements	Percent
I have a gadget for remote learning	100
I have special equipment for PESC at home	36.0
I have already used software for PESC before	17.4
I have equipment for virtual training	4.2

As can be seen from the table, one hundred percent of students had gadgets for remote learning. About seventeen and a half percent had previous experience using software for PESC, and just over four percent had equipment for virtual training.

2. At the beginning of the first semester, a special section was included in the curriculum for the discipline "Informatics". This section aims to provide students with the competence to work with the Workout Trainer, the Dartfish video analysis program, to create a self-monitoring diary of physical health using Excel, etc. In addition, those interested were able to receive advice on using programs for virtual training and any others.

If necessary, physical education teachers also received information on the use of any software at the department of information technology and physical and mathematical disciplines.

3. In order to take into account the individual needs and capabilities of students, they were offered four thematic blocks to choose from, which contained different contents of physical exercises for remote classes. The first block contained only exercises that can be performed indoors. The second included classes both indoors and outdoors (in a park, at a stadium, etc.). The third block was designed to perform exercises

exclusively outdoors. Exercises from the fourth block involved the use of virtual reality programs.

Groups for physical education classes were formed according to the block of exercises chosen by the students.

4. For safe monitoring of remote classes during and after physical exercises, students were recommended to use special hardware or hardware built into smartphones and fitness bracelets: pulse oximeter, pressure meter, pedometer, accelerometer.

During the exercises, it was recommended to make a separate video recording of the training, if it was not included in the program used by the student. After recording and watching the execution of various exercises and their combinations, it was possible to get an evaluation of the performance not only by the teacher, but also to participate in an online tournament. The final control of the performance of the exercises according to the normative indicators also took place on the basis of the video recordings provided by the student, which the student attached to the thematic area in Google Classroom.

4.3 Indicators

At the end of the academic year, a student survey was conducted using two sets of questions. Based on the results of the answers to the first group of questions, Fig. 2 shows how students who chose different blocks of exercises evaluate their physical condition and are motivated to engage in physical education.

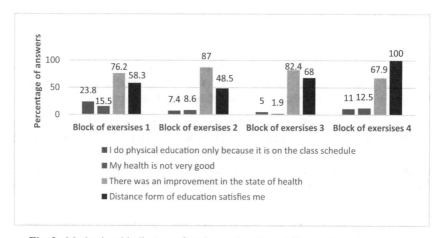

Fig. 2. Motivational indicators of students who chose different blocks of exercises.

When answering the questions of the second group, each student presented indicators of compliance from 0 to 100 points. The designation of the A_i indicator, the questions and the percentage of respondents for whom the A_i indicator had a value greater than 50 are shown in Table 3.

Table 3. The second group of questions.

Indicator A_i	Question	Percent
A_1	To what extent are you satisfied with the remote form of training in physical education?	56.3
A_2	To what extent do you enjoy physical education and sports?	74.3
A_3	Was the time allocated for communication with the teacher sufficient for the correct mastering of physical exercises?	83.9
A_4	To what extent have you mastered the exercise technique?	62.3
A_5	How high-quality and user-friendly was the software used during remote learning?	76.1

Correlation analysis was used to establish the relationship between conditions and results of remote learning. Indicator A_1 was taken as a correlative response (the question "Are you satisfied with the remote form of training in physical education?"). The relationship between it and the indicators of all other answers was established using the stepwise regression method. A statistically significant relationship was established with indicator A_3 (characterizes the sufficiency of time for communication with the teacher) and A_5 (characterizes the ease of use of the software).

A regression equation with a positive correlation was obtained for both indicators:

$$A_1 = 3.61 + 0.35A_3 + 0.42A_5 + 0.11A_4 \tag{1}$$

with the coefficient of determination $R^2 = 0.69$.

There was also a weak correlation of the A_1 indicator with the A_4 indicator (characterizing the mastering of the exercise technique) at a significance level of 5%. Other indicators for determining A_1 were insignificant. Based on the value of the coefficient of determination of 0.69, it can be concluded that 69% of satisfaction with remote learning is due to sufficient communication with the teacher, ease of use of the software, and mastering the exercise technique. The remaining 31% of the response change is determined by other factors not covered by the questions in Table 3.

5 Discussion

A positive answer to the question "Are you satisfied with the remote form of training in physical education?" was provided by more than 55 percent of the students (Table 3). This significantly exceeds the data of other studies on the satisfaction of respondents with remote learning, which is usually limited to 30–40 percent [4]. Therefore, their authors note the low effectiveness of replacing offline training in physical education with a remote form. At the same time, the proposed L&D model of information technology application leads to a significant increase in satisfaction with remote learning.

The analysis of satisfaction with remote learning based on motivational indicators and the block of physical exercises chosen by students showed the following. The most numerous was the group of students who had indoor training software (the first block of exercises). Satisfaction with remote learning, among them, corresponded to the average percentage for all groups (Table 3, Fig. 2).

The best dynamics with the improvement of the state of health occurred among students who used the software both indoors and outdoors (the second block of exercises), and their satisfaction with the remote learning was almost fifty percent (Fig. 2). The least convenient remote learning was for students who preferred classes outside (the third block of exercises) (Fig. 2). It should be noted that in the case of outdoor classes, it is not very convenient to use software and means of video recording of classes.

Analysis of the data in Tables 2 and 3 and Fig. 2 shows that the fourth block of exercises using virtual reality programs is chosen by more students than the number of owners of virtual training equipment. The motivation for additional search for such equipment is obvious. In addition, students who chose the fourth block were the most motivated to use remote learning (100 percent). However, they have the highest rate of change in physical condition for the worse. This indicates an insufficient level of students' ability to use software for virtual training and the need to ensure a deeper mastery of the software in the future.

At the same time, among students of all groups, the percentage of respondents with a high assessment of the quality and convenience of the software used during remote learning was 76.1% (Table 3, indicator A_5). This is a good indicator of software quality. In addition, the use of the model provided a high indicator (83.9%) of the sufficiency of the time allotted to the student for communication with the teacher (Table 3, indicator A_3). According to the results of the correlation analysis, these two indicators were the main ones for ensuring satisfaction with remote learning (expression 1).

6   Conclusions

Regardless of the using location of software (indoors or outdoors), the use of special software (with virtual reality programs) or not, the L&D model of information technology application provided the following.

1. Created a system of informational and technological support for the remote educational process in physical education and support for the physical health of students.
2. Adjusted the educational service to the individual interests and needs of students with the choice of conditions for conducting classes, the maximum pace of mastering and processing of physical exercises.
3. Provided enough time to communicate with the teacher to master physical exercises.
4. Gave the teacher an opportunity to receive information about the physical activity of each student and track the dynamics of his/her state of health.
5. Increased the level of satisfaction with remote learning in physical education.

Further development of the proposed model involves the creation of an online platform for the exchange between educational institutions of practical information on achievements and progressive developments in the methodology of teaching physical education.

References

1. Suryansah, H., Suryadi, L.E.: Use of information technology media on physical activities of students. J. Phys. Conf. Ser. **1539**, 012053 (2020)
2. Varga, A., Révész, L.: Impact of applying information and communication technology tools in physical education classes. Informatics **10**(1), 20–24 (2023)
3. Li, L., Zhang, L., Zhang, S.: Using artificial intelligence for the construction of university physical training and teaching systems. J. Healthc. Eng. **2023**, 1–10 (2021)
4. Rutkauskaite, R., Koreivaite, M., Karanauskiene, D., Mieziene, B.: Students' skills and experiences using information and communication technologies in remote physical education lessons. Sustainability **14**, 15949 (2022)
5. Filenko, L., et al.: Teaching and learning informatization at the universities of physical culture. J. Phys. Educ. Sport. **17**(4), 2454–2461 (2017)
6. Stoicescu, M., Stanescu, M.: Social media as a learning tool in physical education and sports area. In: The 14th International Scientific Conference eLearning and Software for Education, Bucharest, 19-20 April, vol. 3, pp. 346–353 (2018)
7. Chang, K.E., Zhang, J., Huang, Y.S., Liu, T.C., Sung, Y.T.: Applying augmented reality in physical education on motor skills learning. Interact. Learn. Environ. **28**(6), 1–9 (2020)
8. Huang, C.H., Chin, S.L., Hsin, L.H., Hung, J.C., Yu, Y.P.: A web-based e-learning platform for physical education. J. Netw. **6**(5), 721–727 (2011)
9. Norms for performing physical training exercises in 2023. https://ldubgd.edu.ua/abituriientu/normativi-vikonannya-vprav-z-fizichnoyi-pidgotovki. Accessed 22 May 2023
10. New Zealand's leading teaching and research university in Sport and Exercise Sciences and Physical Education. https://www.otago.ac.nz/sopeses/index.html. Accessed 11 Jun 2023
11. Putrantoa, J.S., Heriyantoa, J., Said, A., Kurniawan, A: Implementation of virtual reality technology for sports education and training: systematic literature review. Procedia Comput. Sci. **216**, 293–300 (2023)
12. Lee, H.T., Kim, Y.S.: The effect of sports VR training for improving human body composition. EURASIP J. Image Video Process. **148**, 1–5 (2018)
13. Yang, Y.: The innovation of college physical training based on computer virtual reality technology. J. Discrete Math. Sci. Crypt. **21**(6), 1275–1280 (2018)
14. Petri, K., et al.: Training using virtual reality improves response behavior in karate kumite. Sports Eng. **22**(1), 1–12 (2019)
15. Imas, Y.V., et al.: Modern approaches to the problem of values' formation of students' healthy lifestyle in the course of physical training. Phys. Educ. Students. **22**(4), 182–189 (2018)
16. Buchheit, M., Gray, A., Morin, J.B.: Assessing stride variables and vertical stiffness with GPS-Embedded accelerometers: preliminary insights for the monitoring of neuromuscular fatigue on the field. J. Sports Sci. Med. **14**, 698–701 (2018)
17. Litvinova, M., Andrieieva, N., Zavodyannyi, V., Loi S., Shtanko, O.: Application of multiple correlation analysis method to modeling the physical properties of crystals (on the example of gallium arsenide). Eur. J. Enterp. Technol. **6**(12(102)), 39–45 (2019)
18. Litvinova, M., Dudchenko, O., Shtanko, O.: Application of logistic concept for the organization of small academic groups training in higher education institutions. Naukovyi visnyk Natsionalnoho hirnychoho universytetu. **2**, 174–179 (2022)
19. Dudchenko, O., Litvinova, M., Shtanko, O., Karpova, S.: Using the technical experiment in the computer simulation training for prospecting software engineers. Int. J. Comput. **19**(2), 216–223 (2020)

Maximising Attendance in Higher Education: How AI and Gamification Strategies Can Boost Student Engagement and Participation

Viktoriya Limonova[1]([✉]) [iD], Arnaldo Manuel Pinto dos Santos[1,2] [iD],
José Henrique Pereira São Mamede[1] [iD], and Vítor Manuel de Jesus Filipe[2,3] [iD]

[1] Universidade Aberta, 1269-001 Lisboa, Portugal
2202373@estudante.uab.pt, {arnaldo.santos,jose.mamede}@uab.pt
[2] INESC TEC, 4200-465 Porto, Portugal
vfilipe@utad.pt
[3] School of Science and Technology, Universidade de Trás-Os-Montes E Alto Douro, 5000-801
Vila Real, Portugal

Abstract. The decline in student attendance and engagement in Higher Education (HE) is a pressing concern for educational institutions worldwide. Traditional lecture-style teaching is no longer effective, and students often become disinterested and miss classes, impeding their academic progress. While Gamification has improved learning outcomes, the integration of Artificial Intelligence (AI) has the potential to revolutionise the educational experience. The combination of AI and Gamification offers numerous research opportunities and paves the way for updated academic approaches to increase student engagement and attendance. Extensive research has been conducted to uncover the correlation between student attendance and engagement in HE. Studies consistently reveal that regular attendance leads to better academic performance. On the other hand, absenteeism can lead to disengagement and poor academic performance, stunting a student's growth and success. This position paper proposes integrating Gamification and AI to improve attendance and engagement. The approach involves incorporating game-like elements into the learning process to make it more interactive and rewarding. AI-powered tools can track student progress and provide personalised feedback, motivating students to stay engaged. This approach fosters a more engaging and fruitful educational journey, leading to better learning outcomes. This position paper will inspire further research in AI-Gamification integration, leading to innovative teaching methods that enhance student engagement and attendance in HE.

Keywords: Student Attendance · Student Engagement · Higher Education · Gamification and Artificial Intelligence

1 Introduction

Reduced attendance in Higher Education (HE) has been subject to extensive research by academics. Empirical evidence suggests a direct correlation between consistent attendance and enhanced learning experiences, highlighting concerns about the detrimental effects of low attendance on student engagement and performance [1]. Consistent attendance is essential for intellectual engagement, enabling students to comprehend the presented material [2]. Student absenteeism and lack of motivation are prevalent issues that impede the educational process and limit students' career opportunities [3, 4]. Low attendance can pose significant challenges to HE institutions [5]. Student engagement in HE involves activities, behaviours, and attitudes crucial for academic, personal, and social development [6]. HE institutions should create an interactive learning environment that fosters meaningful interactions to support student development and promote academic achievement [7]. Recent studies have found a correlation between HE attendance and academic performance [8–10]. Active engagement is vital to achieve academic success [9]. How learning platforms and courses are designed and structured is crucial in determining students' paths on their educational journey [10]. Flexible and adaptable teaching strategies are essential for providing a rich educational experience [7]. Technology and modern teaching methods offer better learning opportunities, improving education [11].

This study presents a solution to improve engagement and attendance rates in traditional lecture-based education. By integrating advanced technology with conventional teaching methods, this initiative aims to enhance the lecture experience and facilitate more efficient learning outcomes. With this project, we aim to answer the following research question: How can we solve the lack of student engagement in Higher Education through technology, and how does it affect attendance?

The paper is divided into four chapters, with the first chapter focusing on the relationship between class attendance and student engagement in HE. The second chapter examines the impact of gamification and AI in Education, highlighting their potential to improve student engagement. The third chapter outlines the methodology employed in the research, employing the Design Science Research methodology and explaining the six phases involved. Finally, the last chapter presents a comprehensive discussion and conclusion, addressing all assumptions and limitations while providing details about future extensions of our work and bringing together all the key findings.

2   Preliminary Literature Review

2.1   Gamification in Education

Gamification involves adding game-based elements to non-gaming settings to increase engagement and motivation. This technique turns mundane activities into interactive experiences by introducing a game-like structure that rewards sustained efforts and recognises positive contributions [12]. In Education, Gamification blends enjoyable and intriguing game aspects with structured educational environments. Integrating technology in teaching and learning significantly enhances course participation and engagement [13]. An in-depth survey of recent academic research in art studies highlights Gamification as a critical strategy for amplifying motivation [4, 10].

Kalogiannakis et al. [14], conducted a literature review on gamified science education settings and found that several articles lacked a solid theoretical foundation. Only a few incorporated established models such as intrinsic motivation, engagement and principles for goal-setting and achievement. The study also revealed that game-like elements, such as leaderboards, points, and levels, were often included in gamified science education settings. Meanwhile, a comprehensive study by van Roy and Zaman [15], explored the impact of game design elements in learning environments on fulfilling students' basic psychological needs. Implementing Gamification in educational settings has been found to enhance students' sense of autonomy, competence, and relatedness.

HE institutions are grappling with low student engagement and frequent absenteeism; this issue notably impacts academic performance, particularly in rigorous courses. Effective strategies are required to reignite students' interest and improve outcomes [10]. Gamification has proven to be a highly effective tool in enhancing the learning experience [11]. Studies have consistently shown that implementing Gamification in educational settings significantly impacts students' motivation levels [10, 11]. Incorporating Gamification techniques into attendance tracking can increase student engagement and promote a healthy sense of competition.

2.2 AI's Role in Revolutionizing Gamification in Education

The emergence of AI technology is revolutionising various industries across the globe. AI holds immense promise for tackling some of the most intricate issues faced by humankind and is widely recognised as a symbol of advancement [16]. AI is a revolutionary technology that enables machines to imitate human cognitive abilities, including learning, reasoning, problem-solving, perception, and language comprehension [17].

In Education, instructors need to share their perspectives as we continue to navigate the educational and ethical ramifications of integrating AI into higher learning [18]. AI is an essential tool for enhancing Education through data analysis and Gamification. By identifying behavioural patterns and significant errors, AI allows educators to personalise their approach to each student's unique needs, reinvigorating motivation and engagement [18]. Despite limited research in this area, the novelty of this integration highlights its potential to inspire further academic inquiries and promote modernised strategies. The vast capabilities of AI can lead to customised and interactive learning experiences, ultimately improving student outcomes [18]. AI-augmented gamified systems hold the potential to deliver tailored educational experiences. By continuously monitoring students' interactions and performance, these systems can identify individual strengths, preferences, and areas of challenge [19]. However, a review of AI in Education (AIEd) research by Crompton and Burke [18], revealed some gaps in the lack of research in developing nations, a concentration of studies in specific fields such as language learning, computer science, and engineering, and a growing involvement of education faculties in AIEd research. There is a surging interest in creating game-based or gamified AI educational robotics (AIERs) to promote student participation and improve learning efficacy, tackling the problem of student disengagement. Nonetheless, there exists a conspicuous lack of game-based approaches in AIERs, indicating a promising opportunity for further investigation and advancement [19].

AI has made remarkable progress in Education, especially with the development of Intelligent Tutoring Systems. These systems are known for their adaptability and precision, as they are designed to meet student's unique learning needs and provide personalised and adaptive feedback, allowing for a challenging yet achievable learning experience [18, 20]. The field of Education has been transformed by AI technology, which now offers personalised learning programs tailored to every student's strengths and weaknesses [20].

3 Methodology

This project will employ the Design Science Research (DSR) methodology, which prioritises creating innovative and pragmatic solutions. DSR is ideally suited for identifying critical challenges that span various application domains. Its focus is on the prompt and effective development of solutions that directly address current needs, thereby advancing the field of study [21]. DSR is vital in bridging theoretical knowledge with practical applications, resulting in tangible enhancements in the examined context. According to [21], the DSR methodology consists of six stages, as illustrated in Fig. 1.

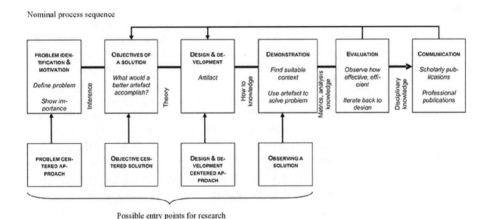

Fig. 1. Design Science Research Model (Adapted from [21])

Table 1 comprehensively maps this study's activities to the various stages of the DSR Methodology. The initial stage, "Problem Identification," is followed by "Objective Definition," which targets enhancing student engagement strategies. The subsequent "Design and Development" phase establishes the project's foundation. "Demonstration" is the next step, testing the artefact's functionality, while "Evaluation" assesses its efficacy. Finally, "Communication" disseminates the findings of the project.

Table 1. Integration of DSR stages in project execution, adapted from [21], with author-specific enhancements regarding the description of each phase.

DSR Stage	Description
Problem Identification	Identification of the main reasons for student absenteeism and reduced interest in HE, with a detailed investigation. This includes a comprehensive review of relevant literature, surveys, and potentially initial interviews with students and educators to collect qualitative and quantitative information on the causes of lack of attendance
Objective Definition	Clear goals with a focus on developing strategies that enhance student engagement. This involves thoroughly investigating the role of Gamification and AI in increasing attendance rates. Objectives should be measurable and address the identified problems
Design and Development	Development of a prototype by synthesising insights from a systematic literature review. Convene a panel of Gamification, AI, and HE experts to discuss initial findings and gather feedback. Utilise this expert input and data from the literature to iterate and refine the design, resulting in a second, more polished version of the artefact
Demonstration	Practical implementation of the artefact in a HE setting to showcase its functionality in a real-world scenario, it is essential to implement it in a suitable environment that represents the target audience and specific needs. Establishing a timeline with a detailed plan, including potential challenges, is essential for the implementation phase
Evaluation	Comprehensive evaluation of the artefact using qualitative and quantitative methods. Data collected should be analysed to assess the artefact's effectiveness in addressing the identified problem. Evaluation should also include a feedback loop in the design phase to make necessary improvements based on the findings
Communication	Writing scientific articles to disseminate research results and contribute to the academic and educational community

4   Discussion and Conclusion

Integrating Gamification and AI in Education holds immense potential to transform the learning experience. These technologies can significantly enhance student motivation and engagement by incorporating gaming elements into academic curricula and leveraging AI for personalised learning experiences. However, adopting these technologies poses several challenges, including infrastructure requirements, educator training, data privacy concerns, and potential biases. Higher education institutions, facing declining attendance and student interest, can benefit from the unique blend of Gamification principles and AI advancements.

This position paper is predicated on several crucial assumptions that underpin its framework and findings. To begin with, the viability of the suggested AI and gamification approach hinges upon sufficient technological infrastructure within educational institutions. It presupposes that there are ample digital resources and connectivity to seamlessly incorporate AI systems and gamified learning platforms into current educational models. Additionally, the efficacy of these measures is contingent upon a certain degree of digital literacy among both students and educators.

However, the study acknowledges limitations, including potential biases in AI algorithms and the scalability of Gamification strategies. Moreover, the lack of long-term empirical studies on these technologies underscores the need for further research to understand their long-term implications in Education. One area of focus is the advancement of more sophisticated AI algorithms that can be tailored to cater to distinct learning styles and needs. Another potential research area is exploring the impact and applicability of Gamification strategies across various academic disciplines. While this study primarily addresses general engagement in HE, future research could delve into the specifics of Gamification in different subjects such as science, humanities, and arts. Examining the integration of gamified elements into different subject areas will provide deeper insights into the versatility and adaptability of Gamification in Education. Longitudinal studies represent a crucial research direction for the future. The long-term effects of AI and Gamification on student engagement, retention, and academic performance are still relatively unexplored. Future studies should track these outcomes over extended periods to evaluate the sustainability and long-term benefits of AI and Gamification integration in learning. This will provide valuable data on the lasting effects of these technological interventions in educational settings.

In summary, this position paper has underscored the transformative potential of integrating AI and Gamification into HE, a paradigm shift to elevate student engagement and motivation. The paper illuminates the path forward for these innovative educational strategies while recognising the inherent challenges. It emphasises the need for robust technological infrastructure, digital literacy, and an ongoing commitment to address and refine AI and gamification methodologies. The proposed research aims to enhance the education system for better engagement, adaptation, and inclusivity, ensuring that the educational landscape remains responsive to the needs and aspirations of the digital age.

References

1. Ancheta, R.F., Daniel, D., Ahmad, R.: Effect of class attendance on academic performance. European J. Educ. Stud. **8** 9, (2021). https://doi.org/10.46827/ejes.v8i9.3887
2. Al-Tameemi, R., et al.: Determinants of poor academic performance among undergraduate students-A systematic literature review. Int. J. Educ. Res. Open **4**(4), 100232 (2023). https://doi.org/10.1016/j.ijedro.2023.100232
3. Rissanen, A.: Student engagement in large classroom: the effect on grades, attendance, and student experiences in an undergraduate biology course. Can J Sci Math Techn **18**, 136–153 (2018). https://doi.org/10.1007/s42330-018-0015-2
4. Pinter, R., Čisar, S.M., Balogh, Z., Manojlović, H.: Enhancing higher education student class attendance through gamification. Acta Polytechnica Hungarica **17**(2), 13–33 (2020). https://doi.org/10.12700/APH.17.2.2020.2.2

5. Farooq, F., Rathore, F.A., Mansoor, S.N.: Challenges of online medical education in pakistan during COVID-19 pandemic. J. Coll. Physicians Surg. Pak. **30**(6), 67–69 (2020)
6. Amerstorfer, C.M., Freiin, C., von Münster-Kistner,: Student perceptions of academic engagement and student-teacher relationships in problem-based learning. Front. Psychol. **12**, 713057 (2021). https://doi.org/10.3389/fpsyg.2021.713057
7. Ferrer, J., Ringer, A., Saville, K., et al.: Students' motivation and engagement in HE: the importance of attitude to online learning. High. Educ. **83**, 317–338 (2022). https://doi.org/10.1007/s10734-020-00657-5
8. Büchele, S.: Evaluating the link between attendance and performance in higher education: the role of classroom engagement dimensions. Assess. Eval. High. Educ. **46**(1), 132–150 (2021). https://doi.org/10.1080/02602938.2020.1754330
9. Fadelelmoula, T.: The impact of class attendance on student performance. Int. Res. J. Med. Med. Sci. **6**, 47–49 (2018). https://doi.org/10.30918/IRJMMS.62.18.021
10. Barna, B., Fodor, S.: An empirical study on the use of gamification on IT. Int. Conf. Interact. Collab. Learn. **715**, 684–692 (2017). https://doi.org/10.1007/978-3-319-73210-7_80
11. Subhash, S., Cudney, E.A.: Gamified learning in higher education: a systematic review of the literature. Comput. Hum. Behav. **87**, 92–206 (2018)
12. Deterding, S., Khaled, R., Nacke, L., Dixon, D.: Gamification: toward a definition. In CHI 2011 Gamification Workshop Proceedings, pp. 12–15 (2011)
13. Bond, M., Buntins, K., Bedenlier, S., Zawacki-Richter, O., Kerres, M.: Mapping research in student engagement and educational technology in higher education: a systematic evidence map. Int. J. Educ. Technol. Higher Educ., **17**(1), 2 (2020)
14. Kalogiannakis, M., Papadakis, S., and Zourmpakis, A.-I..: Gamification in science education. a systematic review of the literature. Educ. Sci. **11**(1), 22 (2021)
15. van Roy, R., Zaman, B.: Unravelling the ambivalent motivational power of gamification: a basic psychological needs perspective. Int. J. of Human-Comput. Stud. **127**, 38–50 (2019). https://doi.org/10.1016/j.ijhcs.2018.04.009
16. Tai, M.-T.: The impact of artificial intelligence on human society and bioethics. Tzu Chi Med. J. **32**(4), 339 (2020). https://doi.org/10.4103/tcmj.tcmj_71_20
17. Tahiru, F.: AI in education: a systematic literature review. J. Cases Inform. Technology **23**(1), 1–20 (2021). https://doi.org/10.4018/JCIT.2021010101
18. Crompton, H., Burke, D.: Artificial intelligence in higher education: the state of the field. Int. J. Educ. Technol. High. Educ. **20**(1), 22 (2023). https://doi.org/10.1186/s41239-023-00392-8
19. Yang, Q.-F., Lian, L.-W., Zhao, J.-H.: Developing a gamified artificial intelligence educational robot to promote learning effectiveness and behavior in laboratory safety courses for undergraduate students. Int. J. Educ. Technol. High. Educ. **20**(1), 18 (2023). https://doi.org/10.1186/s41239-023-00391-9
20. Tapalova, O., Zhiyenbayeva, N., Gura, D.: Artificial Intelligence in Education: AIEd for Personalised Learning Pathways. Electron. J. e-Learn. **20**(5), 639–653 (2022)
21. Peffers, K., Tuunanen, T., Rothenberger, A., Chatterjee, S.: A design science research methodology for information systems research. J. of Manag. Info. Sys. **24**, 45–77 (2007)

Psychometric Properties of a Scale Designed to Assess Satisfaction and Continued Intention to Utilize E-learning Among Nutrition Sciences Students

Leandro Oliveira[1]([✉]) [ID] and Eduardo Luís Cardoso[2] [ID]

[1] CBIOS - Universidade Lusófona's Research Center for Biosciences & Health Technologies, Campo Grande 376, 1749-024 Lisboa, Portugal
leandroliveira.nut@gmail.com
[2] Universidade Católica Portuguesa, CBQF - Centro de Biotecnologia e Química Fina – Laboratório Associado, Escola Superior de Biotecnologia, Rua de Diogo Botelho, 1327 4169-005 Porto, Portugal

Abstract. E-learning uses information and communication technologies to facilitate access to online teaching resources and encourage collaborative environments among students. However, the abrupt transition to e-learning is a challenge in clinical health courses, like nutrition sciences courses, which rely on hands-on training. Several studies indicate that e-learning can be more effective than traditional teaching methods. However, studies and validated tools for evaluating satisfaction and intention to use e-learning are scarce. The main objective of this study is to translate and validate a questionnaire designed to assess the factors that affect the satisfaction and continued intention of higher education students, with a special focus on Nutrition Sciences students, in the e-learning context. An online questionnaire was created and applied between February 2021 and October 2023, distributed during classes. The questionnaire included questions about sociodemographic characteristics and assessed expectations, attitudes, intentions, and the use of distance learning, using a 5-point Likert scale. The analysis showed that the questionnaire is reliable, with a high Cronbach's alpha coefficient ($\alpha = 0.970$) and valid, with good adequacy in exploratory factor analysis (KMO = 0.931). Furthermore, analysis of the scree plot suggested a one-factor solution, where a single factor explained 63.9% of the total variance. This confirms the effectiveness of the questionnaire to assess students' satisfaction and continued intention to use e-learning.

Keywords: Psychometric Properties · Scale · E-learning · Students · Higher Education

1 Introduction

E-learning, defined as the use of information and communication technology (ICT) to improve the quality of education, is an approach that involves the use of electronic technologies and devices in teaching [1]. Literature reviews highlight the advantages and

Á. Rocha et al. (Eds.): WorldCIST 2024, LNNS 988, pp. 71–79, 2024.
https://doi.org/10.1007/978-3-031-60224-5_8

disadvantages of online learning [1, 2]. Previous studies, such as that by Sharif-Nia et al. (2023), reported that students' computer literacy plays a significant role in predicting e-learning acceptance. Furthermore, it was found that computer competence acts as a mediator in the relationship between the role of instructors and course content and students' acceptance of e-learning [3]. More recent research in the field of nutrition sciences in Iran indicates that most students see e-learning as an opportunity to overcome academic challenges, although they do not believe it is as effective as face-to-face teaching [4]. In another study, Salahshouri et al. [5] identified critical factors that influence the adoption of e-learning by students, highlighting aspects such as comparative advantages, observability, and flexibility.

Furthermore, Salloum et al. [6] found that innovation, quality, trust and knowledge sharing play an important role in the acceptance of the e-learning system by students. A cross-sectional study conducted with Italian university students following the COVID-19 pandemic highlighted that sociability, stress, quality of life and coping strategies were crucial factors influencing students' satisfaction with e-learning [7].

Since students represent the future generation of nutritionists, it is extremely important to investigate their opinions regarding the online learning approach. Unfortunately, in Portugal, there is a lack of knowledge about the acceptance of e-learning. Therefore, the aim of this study is the translation and validation of a scale that assesses the factors that influence the satisfaction and continued intention of higher education students, in particular, those studying Nutrition Sciences, in relation to e-learning.

The paper is organized as follows: Sect. 2 outlines the methodology used in the translation and validation process; Sect. 3 presents the results of the scale validation, highlighting key findings and statistical analyses; Sect. 4 discusses the implications of the study results and their relevance to the field of e-learning and higher education; finally, Sect. 5 offers insights into possible future directions for research based on the results of this study.

2 Methodology

An online questionnaire was created and applied from February 2021 to October 2023. This questionnaire was applied annually, covering the beginning and end of one academic semester per year (except for the last one, which only covers the beginning of the semester). Its dissemination took place during classes, being aimed at students enrolled in the Degree in Nutrition Sciences. Questions were included for sociodemographic characterization (Sect. 1), as well as to assess expectations, attitudes, intentions, and use of distance learning (Sect. 2).

The second section of the questionnaire was constructed based on the scale developed in the study by Rajeh et al. [8], which aimed to identify the factors that influence the satisfaction and continued intention of medical and dental students regarding e-learning. This scale was translated from English to Portuguese (Annex 1) and then translated back into English to check whether the meaning of the items remained unchanged.

The scale consisted of 20 items evaluated on a 5-point Likert-type response scale, where participants selected the level that best reflected their perception regarding each item, ranging from 0 (strongly disagree) to 4 (strongly agree). Items include statements related to topics, such as: expectation (items 1, 2, and 3), attitude (items 4, 5, and 6), subjective norms (items 7, 8, and 9), perceived behavioral control (items 10, 11, and 12), confirmation (items 13, 14, and 15), satisfaction (items 16 and 17), and intention to use (items 18, 19, and 20). The final score was calculated by adding the scores assigned to each item on the scale. The total score, which resulted from the sum of the scores for all items, ranged from 0 to 80. Higher scores, both in each individual item and in total, indicated more favorable perceptions regarding the continued intention to use e-learning.

This study was approved by the Ethics Committee of the School of Health Sciences and Technologies of Universidade Lusófona (P28–23).

Statistical analysis was performed using the IBM SPSS Statistics software, version 26.0, on the Windows operating system. To compare the responses to each item between the different sexes, the Mann-Whitney test was applied. Due to the absence of statistically significant differences, the analysis was conducted considering the sample as a whole. The internal consistency of the scale was assessed using Cronbach's alpha coefficient, and items with item-total correlations below 0.2 were excluded. The scale was subjected to a factor analysis using the principal components extraction method, without applying rotation. The factor analysis models were validated using the Kaiser-Meyer-Olkin (KMO) sampling adequacy test and the Bartlett test. The scree plot method was used to determine the number of components to retain in the factor analysis [9]. The null hypothesis was rejected when the critical significance value (p) was below 0.05.

3 Results

The sample used in this study included 72 students from the Nutrition Sciences course at Universidade Lusófona, located in Lisbon. The majority of participants were female, representing 75.0% of the total, with an average age of 22.3 years old (standard deviation: 6.3). Around 90% were of Portuguese nationality, 4.3% Angolan, and 2.8% Brazilian.

An assessment of the psychometric properties of the questionnaire was carried out, including construct validity and internal consistency (reliability), as well a principal component analysis. A final solution consisting of 20 items was obtained – Table 1.

Cronbach's alpha coefficient ($\alpha = 0.970$) indicates that the scale has excellent internal consistency. Furthermore, the results of the Kaiser-Meyer-Olkin test, which obtained a value of 0.931, and the Bartlett sphericity test ($p < 0.001$) confirm the adequacy of the factor analysis. When performing factor analysis using the principal components extraction method, a three-factor solution was obtained, each with an eigenvalue greater than 1. These factors encompassed the 20 original items of the scale and explained 83.9% of the total variance. All items presented factor loadings greater than 0.3, with communalities varying between 0.627 and 0.847. Although the factor analysis initially generated three components with eigenvalues greater than 1, analysis of the scree plot (Fig. 1) indicated that a single-factor solution was more appropriate, where the single latent factor explained 63.9% of the total variance.

Table 1. Reliability and principal component analysis.

	Corrected item-total correlation	Cronbach's α if the item is excluded	Component Matrix		
			C1	C2	C3
1. Using e-learning can improve my learning performance	0.824	0.968	0.845	−0.201	−0.092
2. Using e-learning can increase my learning effectiveness	0.850	0.968	0.870	−0.133	−0.138
3. I find e-learning to be useful to me	0.817	0.968	0.839	−0.157	−0.249
4. Using e-learning is a good idea	0.834	0.968	0.854	−0.212	−0.214
5. I like using e-learning	0.843	0.968	0.862	−0.091	−0.220
6. It is desirable to use e-learning	0.848	0.968	0.865	−0.186	0.034
7. People important to me support my use of e-learning	0.763	0.969	0.790	−0.433	−0.001
8. People who influence me think that I should use e-learning	0.762	0.969	0.788	−0.469	0.037
9. People whose opinion I value prefer that I should use e-learning	0.817	0.968	0.838	−0.375	0.059
10. Using e-learning system was entirely within my control	0.729	0.969	0.754	−0.154	0.408
11. I had the resources. Knowledge. And ability to use e-learning	0.612	0.970	0.639	−0.018	0.618
12. I would be able to use the e-learning system well for learning process	0.609	0.970	0.637	0.185	0.624

(*continued*)

Table 1. (*continued*)

	Corrected item-total correlation	Cronbach's α if the item is excluded	Component Matrix		
			C1	C2	C3
13. My experience using the e-learning system was better than I expected	0.761	0.969	0.787	0.303	−0.006
14. The service level provided by the e-learning system was better than I expected	0.703	0.969	0.735	0.290	−0.048
15. The e-learning systems can meet demands in excess of what I required for the service	0.818	0.968	0.840	0.246	−0.164
16. I am satisfied with the performance of e-learning	0.778	0.968	0.803	0.325	0.068
17. I am pleased with the experience of using e-learning	0.816	0.968	0.838	0.321	−0.015
18. My decision to use e-learning system on a regular basis in the future	0.684	0.969	0.716	0.414	−0.128
19. I will frequently use the e-learning system in the future	0.732	0.969	0.760	0.416	−0.076
20. I will strongly recommend that others use it	0.854	0.968	0.873	0.084	−0.164
Cronbach's α	0.970				
Eigenvalue			12.788	1.574	1.210
Variance (%)			63.9	7.9	6.1
Kaiser-Meyer-Olkin			0.931		
Bartlett (p)			< 0001		

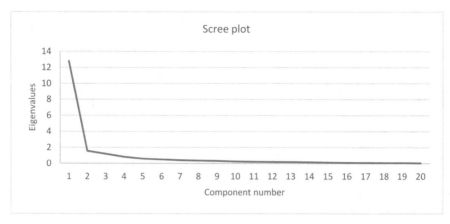

Fig. 1. Scree plot.

4 Discussion

The main purpose of this study was the translation and analysis of the psychometric properties of a scale designed to assess satisfaction and continued intention to use e-learning. A sample of undergraduate students in nutritional sciences was used, and statistical tests were carried out analyzing the reliability and principal components of the scale.

The assessment of item homogeneity was conducted by calculating Cronbach's alpha coefficient ($\alpha = 0.970$), which demonstrated solid internal consistency in the questionnaire. Exploratory factor analysis revealed a notable interrelationship between the variables, as indicated by the Kaiser-Meyer-Olkin (KMO) test value of 0.931, which is considered satisfactory for factor analysis [9]. Furthermore, when considering the scree plot, it was suggested to extract only one latent factor, which managed to explain around 63.9% of the total variability observed in the data. These results align harmoniously with the findings of the original study from which the questionnaire questions were translated [8]. The developed questionnaire proved to be highly practical and convenient for application to students, since the average completion time did not exceed 10 min and there were no reports of significant difficulties during the application process.

This work has some limitations. One of them is its cross-sectional design, which does not make it possible to extrapolate the results to periods beyond the observation period. Furthermore, the sample used is relatively small and focused specifically on students studying Nutrition Sciences, which may restrict the ability to generalize the results to other populations. However, it is suggested that future studies be carried out capable of establishing normative data, considering sociodemographic variables. This approach would provide a solid basis for the interpretation and contextualization of results, enriching the understanding of students' perceptions regarding e-learning.

5 Conclusion

This study made it possible to translate and validate a scale that demonstrated appropriate properties for assessing satisfaction and continued intention to use e-learning among higher education students, in particular, Nutritional Sciences students. This instrument can be used, with the necessary adaptations, to evaluate satisfaction in online courses aimed at students, as well as to measure expectations, attitudes, intentions, and the continued use of distance learning by students. Furthermore, it is important to highlight some future directions that can be explored based on the results of this study. Firstly, it is recommended that additional research be carried out to validate the scale in different educational contexts and with broader samples. Additionally, it would be valuable to conduct longitudinal studies to monitor the evolution of satisfaction and intention to use e-learning over time. This would allow for a deeper understanding of the dynamics of these indicators and provide valuable insights to improve the quality of online teaching. These future directions can contribute to the continued development of effective distance learning strategies, promoting a more enriching and satisfying educational experience for students.

Funding.. This work was co-financed by the European Union - ERASMUS + Program. Through the SMARTI project (617489-EPP-1-2020-1-DE-EPPKA2-CBHE-JP) and Fundação para a Ciência e a Tecnologia (FCT) through projects DOI 10.54499/UIDP/04567/2020 and DOI 10.54499/UIDB/04567/2020 to CBIOS.

Annex

Annex 1. Scale items and constructs in English (original) and Portuguese (translation).

English	Portuguese
Expectation	*Expectativa*
1. Using e-learning can improve my learning performance	1. Através do ensino à distância posso melhorar a minha aprendizagem
2. Using e-learning can increase my learning effectiveness	2. Através do ensino a distância posso aumentar a eficácia da minha aprendizagem
3. I find e-learning to be useful to me	3. Penso que o ensino a distância seja útil para mim
Attitude	*Attitudes*
4. Using e-learning is a good idea	4. Penso que utilizar o ensino a distância seja bom
5. I like using e-learning	5. Eu gosto de utilizar o ensino a distância
6. It is desirable to use e-learning	6. É desejável o recurso ao ensino a distância

(*continued*)

78   L. Oliveira and E. L. Cardoso

(continued)

English	Portuguese
Subjective norms	*Normas subjetivas*
7. People important to me support my use of e-learning	7. As pessoas que são importantes para mim pensam que é positivo frequentar o ensino a distância
8. People who influence me think that I should use e-learning	8. As pessoas que me influenciam pensam que é positivo frequentar o ensino a distância
9. People whose opinion I value prefer that I should use e-learning	9. As pessoas cuja opinião eu valorizo pensam que é positivo frequentar o ensino a distância
Perceived behavioural control	*Controlo comportamental percebido*
10. Using e-learning system was entirely within my control	10. A utilização do sistema de ensino a distância está totalmente dentro das minhas capacidades
11. I had the resources. Knowledge. And ability to use e-learning	11. Eu tenho os recursos. Conhecimentos e competências para frequentar um curso de ensino a distância
12. I would be able to use the e-learning system well for learning process	12. Eu seria capaz de utilizar uma plataforma de ensino a distância para aprender
Confirmation	*Confirmação*
13. My experience using the e-learning system was better than I expected	13. A minha experiência na utilização de uma plataforma de ensino a distância foi melhor do que eu esperava
14. The service level provided by the e-learning system was better than I expected	14. Os recursos oferecidos pela plataforma de ensino a distância superaram as minhas expetativas
15. The e-learning systems can meet demands in excess of what I required for the service	15. As plataformas de ensino a distância vão de encontro às minhas necessidades de aprendizagem
Satisfaction	*Satisfação*
16. I am satisfied with the performance of e-learning	16. Estou satisfeito com o meu desempenho no ensino a distância
17. I am pleased with the experience of using e-learning	17. Estou satisfeito com a minha experiência da utilização do ensino a distância
Intention to use	*Intenção de utilização*
18. My decision to use e-learning system on a regular basis in the future	18. Tenho intenção de utilizar plataformas de ensino a distância de forma regular no futuro
19. I will frequently use the e-learning system in the future	19. Utilizarei frequentemente plataformas de ensino a distância no futuro
20. I will strongly recommend that others use it	20. Eu recomendo fortemente que outros utilizem plataformas de ensino a distância

References

1. Haleem, A., Javaid, M., Qadri, M.A., Suman, R.: Understanding the role of digital technologies in education: a review. Sustain. Oper. Comput. **3**, 275–285 (2022)
2. Topping, K.J.: Advantages and disadvantages of online and face-to-face peer learning in higher education: a review. Educ. Sci. **13**(4), 326 (2023)
3. Sharif-Nia, H., et al.: E-learning acceptance: the mediating role of student computer competency in the relationship between the instructor and the educational content. Teaching and Learning in Nursing 2023 (2023)
4. Eslamian, G., Khoshnoodifar, M., Malek, S.: Students' perception of e-learning during the Covid-19 pandemic: a survey study of Iranian nutrition science students. BMC Med. Educ. **23**(1), 598 (2023)
5. Salahshouri, A., et al.: The university students' viewpoints on e-learning system during COVID-19 pandemic: the case of Iran. Heliyon **8**(2), e08984 (2022)
6. Salloum, S.A., Al-Emran, M., Shaalan, K., Tarhini, A.: Factors affecting the E-learning acceptance: a case study from UAE. Educ. Inf. Technol. **24**(1), 509–530 (2019)
7. Cofini, V., et al.: E-Learning Satisfaction, Stress, quality of life, and coping: a cross-sectional study in italian university students a year after the COVID-19 Pandemic Began. Int. J. Environ. Res. Public Health **19**(13), 8214 (2022)
8. Rajeh, M.T., et al.: Students' satisfaction and continued intention toward e-learning: a theory-based study. Med. Educ. Online **26**(1), 1961348 (2021)
9. Maroco, J.: Análise estatística com utilização do SPSS, 3.ª edn. Lisboa: Edições Silabo, Lda (2007)

Software Testing in Ecuadorian University Education: A Debt to the Software Industry

Belén Cerón[2], Alex Chicaiza[2], Johana Lamiño[3], and Rolando P. Reyes Ch[1](✉) ⓘ

[1] Universidad Politécnica de Madrid, Campus Montegancedo, Madrid, Spain
rp.reyes@alumnos.upm.es
[2] Universidad de Las Fuerzas Armadas ESPE, Sangolquí 171103, Ecuador
{mbceron,agchicaiza2}@espe.edu.ec
[3] Universidad Internacional de La Rioja, Madrid, Spain

Abstract. The software industry has evolved exponentially in the last 50 years, and with it, software testing has become like a badge of software quality. However, during this evolution universities apparently have not kept pace with the changing and dynamic pace of the software industry. We have been motivated to carry out an exploratory study to analyze the current state of teaching software testing in careers related to Computer Science in Ecuadorian universities. We use a survey as an empirical strategy with a representative population of students in the country. The results are interesting and indicate possible problems in the teaching and culture of use of software testing. Consequently, this deficiency in the correct application of software testing by students is taken to the professional field and therefore to a high-growth industry. It is concluded that software testing in Ecuadorian university education still has deficiencies, making this a debt to be fulfilled by the software industry in Ecuador.

Keywords: Software engineering · software testing · software development · automated tools · survey

1 Introduction

From its beginnings to the present, software development has been fundamental for innovation in society. For this reason, software development is directed by a wide range of paradigms, techniques, and technologies to achieve solutions to different fields of industry such as: medicine, education, and other sciences in which it can be implemented. A confirmation of this has been indicated by Pressman et al. [1] who has mentioned in several editions of his books that, in the last 50 years, software has gone from being considered a solution to a problem or a simple analysis tool to become a high-growth industry [1]. This would not have been achieved without the existence of software testing and Experimental Software Engineering (ISE). That is why now talking about software testing has become essential and important in the industry [12]. Now, if we ask ourselves the reasons why software testing is so important, we could say that there are several reasons, among the most important we have: 1) that software testing is important badges of quality and reliability, and 2) allow the early detection of defects and the reduction of associated risks in software development [2]. Now that we know the importance of software testing, it is important to know its use, application, design, and execution; and this is where the Software Engineer requires skills and knowledge of techniques or tools. It used to be said several decades ago among software development professionals that not every developer could become a tester. This is because a developer simply did not have the necessary knowledge to be able to work on it, especially "junior" developers (mostly students or recent graduates) who do not know software testing in depth [3, 6, 8]; and that the reasons are normally associated with the fact that adequate learning did not exist in universities, perhaps to the lack of importance or the supremacy of theoretical teaching over practice added to the student's low motivation to become acculturated in software quality [3].

With this background, we were motivated to carry out an exploratory study with the objective of knowing the current state of teaching software testing in various careers related to Computer Science in Ecuadorian universities. This is supported by Garousi et al. [4] who expresses the importance of this type of studies due to the broad demand in the industry that seeks software quality. To investigate this question, we propose the application of a survey as an empirical strategy [9] to sample students from various careers related to Computer Science at different levels within Ecuadorian universities. The results obtained show that there is a deficiency in the teaching of software testing due to theoretical approaches over practical ones, lack of motivation for students in the use of software tests in software development projects during their learning.

This article is structured as follows: Section 2 presents the related works. The design of the survey as an empirical research strategy is detailed in Sect. 3. Section 4 presents the execution and results of the study. Finally, in Sect. 5 some brief conclusions and discussion of the study.

2 Background

Pressman et al. [1] in his various editions of his book, he stated something simple but true: that in the last 50 years, software went from being the typical solution to a specialized problem to being a large and powerful industry. In other words, software development companies have prioritized the quality of their processes and software product, and this is where software testing becomes the most important player in the software development industry [2]. Several years ago, software testing became one of the most important and necessary skills within the software industry [5]. Now, knowing, learning, and having skills in software testing is not easy. It requires (apart from experience) the bases and knowledge of at least most of the paradigms, techniques and tools for software verification and validation [12]. This is why the academy plays a very important role in the correct teaching of software testing. Unfortunately, when reviewing some articles from several years ago, this has not been the reality, the same ones that have commented the following:

For example, Timoney [3] detected a learning problem in software testing: such as the existence of various texts, books, articles referring to software testing where their definitions are "very individualistic" and that depended on the point of view of each teacher (laboratory and experimental practice is almost non-existent).

Scatalon et al. [6] in the same sense, carried out his study on graduate students in careers related to computer science in Brazil to know two important aspects. On the one hand, establish the professional profile and know if the graduated students took specialized subjects in software testing; and, on the other hand, to know the gaps that the students had on the topic of software testing. In their results they mention that low percentages of students learned software testing at the university.

Finally, Frezza [8] raises the pressing need that academia must integrate software testing concepts, techniques, and tools during learning so that the student has a greater ability to formulate and execute test plans from test models. The author considers that this integration brings benefits in software testing application skills.

Now, although it is true that the studies presented have been published for some time, the topic is still current and under debate. Elgrably et al. [20] puts the topic on the table and indicates that regardless of the temporality, the teaching of software testing will always be a recognizable and necessary topic, especially its discussion in the organization and depth of teaching. Even Alenezi & Akour [21] recently evaluated software testing skills and how these skills can be enriched during university studies prior to starting their professional jobs, minimizing the misalignment of their academic education with what is needed in the industry [21].

This background motivates us to explore what is happening in careers related to computer science in Ecuadorian universities, especially the teaching of software testing to their students. It is unknown if the same problems mentioned by [3, 6, 8, 20, 21] are occurring. Therefore, we have planned to carry out an exploratory study with three objectives: 1) to know the current state of university students at different levels; 2) know if teachers use theoretical or practical approaches during their teaching (use tools, techniques, etc.); and 3) to know if students at higher levels are motivated to naturally put software testing into practice during the development of their projects.

3   Methodology

To achieve our objectives, we pose our research questions (RQ), conduct the survey design, define the population and sample, and finally describe the threats to validity:

3.1   Research Questions

The research questions are closely tied to the problems indicated in Sect. 2. For this reason, we have posed three research questions for our study:

RQ1: *How is the current state of knowledge related to software testing in basic and intermediate level university students who study careers related to Computer Science prior to taking the subject corresponding to software testing?*

RQ2: *Do teachers of subjects related to programming use theoretical and practical arguments during teaching with the use of automated software testing tools?*

RQ3: *Are upper-level university students motivated to put software testing into practice as part of an indispensable culture in software development in the industry?*

3.2 Survey Design

The survey design uses the survey guides for Experimental Software Engineering proposed by Kitchenham [13] and Punter et al. [16]. The purpose of the survey is to obtain quantitative and qualitative data, so that we can descriptively know important information to answer our RQs. The steps for our survey design are:

1) Construction of the survey: Composed of 20 questions, of which 5 questions are aimed at knowing the characteristics of the respondent. The remaining questions are related to the context of our research. Closed and multiple-choice questions were used. The answers to these questions are rated on the 5-level Likert scale [10] in order not to give rise to erroneous interpretations and/or the provision of an irrelevant or confusing answer. However, in some questions the respondent was also given the opportunity to respond according to her criteria. The survey is planned anonymously and online for a time of at least 6 min (depending on the feedback provided).

2) Evaluation instrument: There were 3 stages of evaluation of the survey. The first stage consisted of review and validation regarding readability, understandability, and potential ambiguities by an external researcher who is not considered in this article. With your feedback, 5 questions were increased, and 2 questions were corrected. In the next stage, the applicability of the survey was piloted with the help of 2 university students, who provided feedback with the correction of 2 questions. In the final stage, we had the collaboration of a Junior QA Analyst with some years of professional experience in software testing (not considered the author of this article) who made suggestions on the readability of 2 questions.

3.3 Population, Sample and Data Collection

The population for this survey is made up of 1,331 basic and intermediate level students of careers related to Computer Science at various universities in Ecuador. The population refers to students who have passed subjects related to programming [11], a prerequisite for starting teaching in software testing. The same survey was not carried out on teachers or professors since the study has an exploratory purpose. However, depending on the results, the study would be expanded to professors.

Data collection was carried out from 04/16/2023 and ended on 07/27/2023. The survey was sent by email to a sample of 188 students from the population. Of this sample, 174 surveys were completed satisfactorily, and 14 surveys were partially completed. The latter were not considered for the study. A response rate of $\frac{174}{188} = 92,55\%$. A high ratio regarding the application of surveys in the field of practice in experimental Software Engineering studies [15].

3.4 Threats to Validity

To describe the threats to the validity of our survey, the four perspectives presented by Wohlin et al. were considered [12]:

Validity of the Conclusion: We believe this threat does not apply because we do not use statistical techniques. The only threat would be the lack of representativeness of the chosen sample. For our study, this threat was mitigated with the representativeness of several students from different careers related to Computer Science.

Internal Validity: The appearance of other influential variables could only appear due to formulation or understanding errors in the survey questions. To mitigate this threat, an introduction and objectives of the study were provided before the respondent began the survey. Likewise, to mitigate possible confusion, the questions were ordered into blocks according to what we want to search for for each research question (RQ). With this, the respondent does not lose the context of the topic addressed. Finally, to guarantee the veracity of the results, confidentiality and anonymity were provided to the respondents.

Construct Validity: A small review of related works was carried out (Sect. 2) to determine the problems and impact of software testing in education and possible problems that professionals had during their software testing practice from their university stage. With this, the survey questions were constructed. Likewise, the survey was evaluated by an external researcher, piloted by students and a Junior QA Analyst with experience in software testing.

External Validity: We believe that the results are externally valid because the selected sample is representative among Computer Science career students from Universities in Ecuador.

4 Execution and Results

4.1 General Characteristics of the Respondents

The 36.37% and 63.63% of those surveyed belong to basic and intermediate levels respectively; of which 85.71% are men and 14.29% are women. Something apparently normal in this type of racing. Regarding students' completion of internships or pre-professional practices, 5.19% confirm having completed them. However, none of the practices they mentioned are related to software development.

4.2 RQ1: *How is the Current State of Knowledge Related to Software Testing in Basic and Intermediate Level University Students Who Study Careers Related to Computer Science Prior to Taking the Subject Corresponding to Software Testing?*

The results related to this RQ are shown in Table 1. We could say that apparently two positions were found. On the one hand, there are students who know something about software testing theory, being 94.59% of respondents. The interesting thing is that the respondents acquired knowledge self-taught and at the university. Only 5.41% mention that they do not know software testing due to a lack of demands from their teachers, or they simply do not know or have not heard about the subject. The types of software testing and its practice, unit tests are highlighted as the most used, followed by functional tests, leaving other types of tests (stress, regression, acceptance, etc.) in apparent ignorance. When asked where and when they put them into practice, according to the taxonomy we carried out, many respondents agree that they used some software testing in software development projects in the Programming subjects [14].

Table 1. Answers related to the research question RQ1

Questions	Answer Options	Total	% Total
Do you know what software testing is?	Yes	70	94.59%
	No	4	5.41%
If you answered YES to the previous question since, what semester have you known about the subject?	1st Semester	44	62.86%
	2nd Semester	18	25.71%
	3rd Semester	7	10.00%
	4th Semester	0	0.00%
	5th Semester	0	0.00%
	6th Semester	1	1.43%
If you answered NO to the previous question, why don't you know about software testing?	Open answer	4	100%
Select the types of software testing that you are familiar with and that you have put into practice	Stress tests	7	9.86%
	Load Test	11	15.49%
	Regression test	2	2.82%
	Functional testing	47	66.20%
	Unit tests	60	84.51%
	Test of performance	24	33.80%
	Acceptance Tests	6	8.45%
	security tests	23	32.39%
	Open-source tests	16	22.54%
	None	4	5.63%
	Others	4	5.63%
Based on what you previously answered, when and where did you put it into practice?	Open answer	67	95.71%
Where did you learn your knowledge of software testing?	By the University	67	95.71%
	in a self-taught way	29	41.43%
	Online courses	12	17.14%
	By classmates	8	11.43%

(*continued*)

Table 1. (*continued*)

Questions	Answer Options	Total	% Total
	For friends	7	10.00%
	By other teachers (no University)	2	2.86%
	None	0	0.00%
	Others	0	0.00%

4.3 RQ2: Do Teachers of Subjects Related to Programming Use Theoretical and Practical Arguments During Teaching with the Use of Automated Software Testing Tools?

The results related to this research question RQ2 are shown in Table 2. It is interesting to know that 71.62% of respondents express the importance of learning software testing during learning programming. This is contrasted with the 81.08% who express the motivation they obtained from their programming subject teachers to start software testing. However, we are concerned about the 18.92% of respondents who mentioned that their programming subject teachers did not provide them with good guidance in software testing. Possibly it is due to a lack of time in class, or simply the teacher did not teach the topic of software testing adequately enough to motivate its completion and execution.

Table 2. Answers related to the research question RQ2.

Questions	Answer Options	Total	% Total
How important do you consider learning to perform software testing?	Very important	53	71.62%
	Important	17	22.97%
	Neutral	4	5.41%
	Less important	0	0.00%
	Nothing important	0	0.00%
Have the professors of programming subjects at the university oriented you on the subject of software testing?	Yes	60	81.08%
	No	14	18.92%
If your answer was YES, what have they told you about it?	Open answer	60	100%
If your answer was NO, could you tell us the reason?	Open answer	14	100%

(*continued*)

Table 2. (*continued*)

Questions	Answer Options	Total	% Total
Has the professor of the programming subject or any software development subject taught you how to use any tool oriented towards software testing (testing tools)?	Yes	25	33.78%
	No	49	66.22%
If your answer was YES, what are the tools you learned?	Open answer	25	100%
If your answer was NO, could you tell us the reason?	Open answer	49	100%

Something also worrying is indicated by the 66.22% of respondents who mention that their programming subject teachers did not teach how to use any tool aimed at software testing (tooltesting). Investigating this information in the open responses, several of the respondents responded that the testing tools that their teachers apparently knew referred to tools for documenting requirements, designing web pages, among others. This gives us to understand that the respondent knows more about the theory of software testing than about its practice and tools. Only 5% of respondents mentioned in their open question that they knew testing tools such as: TestRail, Junit, Jacoco, TestProject, JMeter, among others. This contradiction is possibly due to several typical causes in teaching, such as: an inappropriate design of the academic framework, an inappropriate syllable design, the lack of correct direction of teachers, the excess of theory about practice or simply the student is not motivated to consider software testing as a relevant topic during software development.

4.4 RQ3: *Are Upper-Level University Students Motivated to Put Software Testing into Practice as Part of an Indispensable Culture in Software Development in the Industry?*

The results of this RQ are shown in Table 3. We are struck by the fact that the use of software testing during the development of software projects during learning is quite limited. More worrying is that knowing this particular when we find a current software industry that seeks quality as a competitive advantage [2]. Our assessment is based on the 47.14% of respondents who mentioned that they "occasionally" use software testing during their development projects. However, we are much more concerned to know that 30.00% of respondents mentioned that their teachers of the subjects where software projects are developed also "occasionally" or "almost never" include in their academic evaluation rubric, the application of tests. Software. In this sense, we can infer that apparently there is no quality culture during the software or software product development process. Possibly this is the reason why software testing is not put into practice.

Table 3. Answers related to the research question RQ3.

Questions	Answer Options	Total	% Total
How often do you test software in the software development projects you present at your university?	Always	6	8.57%
	Almost always	18	25.71%
	Occasionally	33	47.14%
	Hardly ever	10	14.29%
	Never	3	4.29%
Do teachers in their grading rubric include the application/use of software tests as part of their grading?	Always	7	10.00%
	Almost always	21	30.00%
	Occasionally	21	30.00%
	Hardly ever	17	24.29%
	Never	4	

5   Conclusions and Discussion

The results obtained for RQ1 give us a slight idea that university students barely know what software testing is in all its practical deployment. The little they know is oriented towards the practicality of unit and functional testing. Possibly this knowledge is acquired self-taught during the learning of subjects related to programming. This is contrasted with the response of students from recent semesters. Since the beginning of the university training of students (basic levels) in the field of software development, there is no culture of application and use of software testing [17], which makes them unaware of the extent of static techniques. And dynamics [18], and possibly the most basic software testing principles would go unnoticed [19]. This could be improved with the proposal of Alenezi, M. & Akour [21] who suggest the periodic review of academic frameworks and the implementation of standards-based courses (e.g., IEEE, others) to reduce the misalignment gap between skills. Acquired in higher education and what the student really requires to guarantee their good insertion in the software industry.

With respect to the results of RQ2 and RQ3, we can generally infer that apparently in Ecuadorian universities the theoretical part in subjects related to software testing is still strongly rooted. Nor can we leave aside those teachers who truly prioritize the practical part, but it seems that they are few. Possibly this is due to an inappropriate design of the syllables of the software testing and programming subjects where they simply give little importance to the topic. Perhaps, because of all the time it takes to teach and achieve skills in students for the practical part. This would allow university students in their training to occasionally carry out software testing in their software development projects and for the culture of software quality to be almost unnoticed during the completion of their projects [19].

Finally, we can mention that in Ecuadorian universities it is not an isolated case, it happens and has happened in several universities worldwide. Elgrably et al. [21] has mentioned it in their study and Scatalon et al. [6] and Timoney [3] have shown us this.

We believe that college students imperatively need to improve their skills in software testing where there is now a high-growth industry. The mission of universities now is to reinvent themselves every day in the teaching of software testing. Even more so when the COVID-19 pandemic left a great lesson for the tester community. Allowing software testing tasks to become a highly globalized job, evolved to such a point that they have become remote jobs for testers worldwide. That is why the Software Engineering community is attentive to this evolution and has issued several recommendations for improving the teaching of software testing with updated reference curriculum proposals based on standards such as indicated by ACM / IEEE [21] or the issuance of updated guides such as the Software Engineering Body of Knowledge (SWEBOK) [21]. The software industry knows for a fact that software quality has become its hallmark that differentiates itself from the competition; and the software industry will be waiting for students with high skills in software testing. However, we can say that in this aspect, at least in Ecuador, universities still owe a debt to the Ecuadorian software industry.

References

1. Roger, S.P., Bruce, R.M.: Software Engineering A Practitioner's Approach, vol. 8, p. 496 (2015)
2. Edgar, S.M., Raquel, M.M., Paula, T.O.: A review to reality of software test automation. Computación y Sistemas **23**(1), 169–183 (2019). https://doi.org/10.13053/CyS-23-1-2782
3. Timoney, J., Brown, S., Ye, D.: Experiences in software testing education: some observations from an international cooperation. Proceedings of the 9th International Conference for Young Computer Scientists ICYCS 2008, pp. 2686–2691 (2008). https://doi.org/10.1109/ICYCS.2008.209
4. Garousi, V., Rainer, A., Lauvås, P., Arcuri, A.: Software-testing education: a systematic literature mapping. J. Syst. Softw. **165** (2020). https://doi.org/10.1016/j.jss.2020.110570
5. Aniche, M., Hermans, F., Van Deursen, A.: Pragmatic sotware testing education. In: SIGCSE 2019 – Proceedings of the 50th ACM Technical Symposium on Computer Science Education, no. Section 3, pp. 414–420 (2019). https://doi.org/10.1145/3287324.3287461
6. Scatalon, L.P., Fioravanti, M.L., Prates, J.M., Garcia, R.E., Barbosa, E.F.: A survey on graduates' curriculum-based knowledge gaps in software testing. Proceedings of the - 2018 IEEE Frontiers in Education Conference (FIE), vol. 2018-Octob, pp. 1–8 (2019). https://doi.org/10.1109/FIE.2018.8658688
7. Edwards, S.H.: Using software testing to move students from trial-and-error to reflection-in-action. In: Proceedings of the SIGCSE Technical Symposium on Computer Science Education, pp. 26–30 (2004). https://doi.org/10.1145/971300.971312
8. Frezza, S.: Integrating testing and design methods for undergraduates: teaching software testing in the context of software design. Proc. Front. Educ. Conf. **3**, 4–7 (2002). https://doi.org/10.1109/fie.2002.1158637
9. Reyes, R.P., Dieste, O., Fonseca, C.E.R.: Pruebas de Hipótesis en experimentos de Ingeniería de Software: Desviaciones respecto a la teoría, CIBSE 2016 - XIX Ibero-American Conf. Softw. Eng., 157–170 (2016). https://doi.org/10.13140/RG.2.1.3056.2164
10. Osinski, I.C., Bruno, A.S.: Categorías de respuesta en escalas tipo likert. Psicothema **10**(3), 623–631 (1998)
11. Informativos, D.: Programa analítico 1, pp. 1–4 (2018)
12. Wohlin, C., Runeson, P., Höst, M., Ohlsson, M.C., Regnell, B., Wesslén, A.: Experimentation in Software Engineering, vol. 9783642290 (2012)

13. Kitchenham, B.A., Pfleeger, S.L.: Personal opinion surveys. In: Shull, F., Singer, J., Sjøberg, D.I.K. (eds.) Guide to Advanced Empirical Software Engineering, pp. 63–92. Springer, London (2008). https://doi.org/10.1007/978-1-84800-044-5_3
14. Unidad De Admisión Y Registro De La Universidad De Las Fuerzas Armadas, p. 27
15. de la Vara, J.L., Borg, M., Wnuk, K., Moonen, L.: Survey on safety evidence change impact analysis in practice : detailed description and analysis (2014)
16. Punter, T., Ciolkowski, M., Freimut, B., John, I.: Conductingon-linesurveys insoftwa-reengineering. In: Empirical Software Engineering, 2003. ISESE 2003. Proceedings. 2003 International Symposiumon, pp.80–88. IEEE (2003)
17. Laplante, P.A., Kassab, M.: What Every Engineer Should Know About Software Engineering. CRC Press (2022)
18. Nwandu, I.C., Odii, J.N., Nwokorie, E.C., Okolie, S.A.: Evaluation of software quality in test-driven development: a perspective of measurement and metrics. Inform. Technol. Comput. Sci. **6**, 13–22 (2022)
19. Aniche, M.: Effective Software Testing: A developer's guide. Simon and Schuster (2022)
20. Elgrably, I., de Oliveira, S., Bezerra, S.R.: A diagnosis on software testing education in the Brazilian Universities. In: 2021 IEEE Frontiers in Education Conference (FIE), pp. 1–8. IEEE (2021)
21. Alenezi, M., Akour, M.: Methodical software testing course in higher education. Int. J. Eng. Pedagogy. **12**, 51–62 (2022)

Examining the Digital Gap: Pre-service Teachers' Needs and Expectations

Ana R. Luís[1]([⊠]) [iD] and Carlos Rodrigues[2] [iD]

[1] CELGA-ILTEC, University of Coimbra, Coimbra, Portugal
aluis@fl.uc.pt
[2] University of Coimbra, Coimbra, Portugal
carlos.rodrigues@uc.pt

Abstract. Portuguese teachers are expected to be digitally proficient, however pre-service training in digital competency is not mandatory in Portugal. Recognizing the need to improve prepare teachers for the digital agenda, a postgraduate training program tailored to the specific needs of pre-service English language teachers was developed. To gather insights from pre-service teachers about the perceived benefits and weaknesses of the course module, a qualitative study was carried out, based on the pre-service teachers' responses to an online survey. The results of this study strongly advocate for the implementation of a targeted digital agenda within pre-service teacher education programs.

Keywords: pre-service teacher · digital skills · digital tools · DigCompEdu

1 Introduction

Modern teaching and learning paradigms have undergone a profound transformation as a result of the incorporation of technology [1, 2]. Educators must be prepared to satisfy the different requirements of children in a technologically advanced society, help them think critically and solve problems digitally, and help them prepare for the difficulties that lie ahead [3]. Teachers need to be proficient in both subject matter and digital competence to successfully incorporate digital resources into their teaching practices [4, 5]. However, the first stage of teacher education is sometimes overlooked, despite the fact that educators must constantly learn and improve throughout their careers [6]. This study presents the results of a post-graduate training program which was developed to enhance the digital skills of pre-service teachers. The idea that pre-service teachers are naturally equipped to handle the technological demands of contemporary education is called into question by our research [7]. The goal of our course module was to close the gap between digital competence development and teacher education by offering training in digital literacy and digital skills [8]. In order for pre-service teachers to succeed in contemporary learning contexts, this research argues that a focused digital agenda is necessary [3].

Digital competence in the context of education is the capacity to evaluate digital information critically, solve problems with digital tools, and communicate effectively using digital media [9]. Although it is frequently linked to students' abilities, digital competency is just as important for teachers. Teachers are essential because they set an example for students on how to use technology wisely, create digitally rich learning environments, and adjust to the ever changing needs of education.

A number of frameworks for digital competence have been created recently to direct the growth of digital skills and competences in the classroom. DigCompEdu [9], a well-known framework that offers a thorough model for digital competency among educators. This framework, unlike others, places equal emphasis on technical proficiency and educational elements, digital citizenship, and technology adaptability [10].

A national program called *Capacitação Digital dos Professores* (Digital Teacher Training Program) has been in progress in Portugal since 2020 when the official action plan for an overall digital transition was launched. The main purpose of the program is to offer teachers specialized training and assistance and develop the skills they need to successfully incorporate digital technology into their teaching.

Despite the common belief that pre-service teachers are digital natives because of their age and familiarity with technology, research has shown that this knowledge does not always translate into educational settings successfully [11]. Very often, pre-service teachers may not have the digital pedagogical expertise needed to effectively incorporate technology into their lesson plans. There is a need for focused training throughout the pre-service education period [12].

The current study aims to investigate the efficacy of a postgraduate course module designed especially for pre-service teachers within the context of a Master's Program for Initial Teacher Education in English Language Teaching. The six-week course module was organized into the six competency areas of DigCompEdu, namely teaching and learning, digital resources, professional engagement, assessment, learner empowerment, and enhancing leaners' digital competency [9]. Two research questions guide this study. The first research question was designed to identify the components of the training which pre-service teachers found most useful. The second research question addresses the course module's weaknesses and aims to find out what pre-service teachers think could be changed or adjusted.

2 Methodology

Through an online survey, a cohort of twelve pre-service teachers (six male and six female) was asked two open-ended questions about the course module: a) "What did you find helpful? Why?"; b) "How could the module be improved? Why?" We conducted a qualitative data analysis on the replies and looked for recurrent themes, which reflected the views of the pre-service teachers. The main themes that came out of the data analysis are listed in Fig. 1, which also show how pre-service teachers (S) are related to each theme.

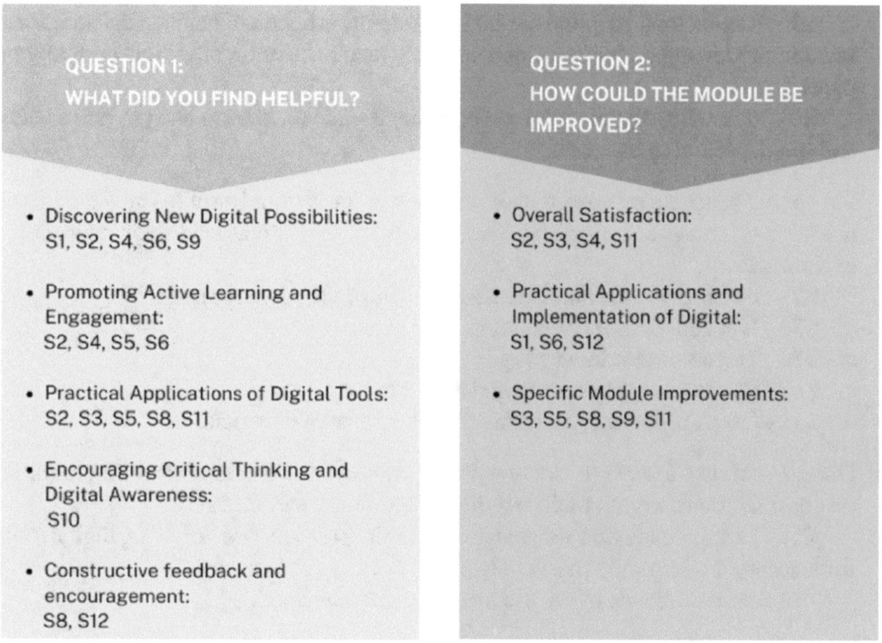

QUESTION 1:
WHAT DID YOU FIND HELPFUL?

- Discovering New Digital Possibilities:
  S1, S2, S4, S6, S9

- Promoting Active Learning and
  Engagement:
  S2, S4, S5, S6

- Practical Applications of Digital Tools:
  S2, S3, S5, S8, S11

- Encouraging Critical Thinking and
  Digital Awareness:
  S10

- Constructive feedback and
  encouragement:
  S8, S12

QUESTION 2:
HOW COULD THE MODULE BE
IMPROVED?

- Overall Satisfaction:
  S2, S3, S4, S11

- Practical Applications and
  Implementation of Digital:
  S1, S6, S12

- Specific Module Improvements:
  S3, S5, S8, S9, S11

Fig. 1. Themes identified in response to Question 1 and Question 2

3   Results

3.1   Question 1

a) Participants expressed excitement about the possibility of coming across a wide range of digital tools and services that they had never used before. They gained new insights into the educational possibilities of technology by learning about new digital tools and resources. They also showed awareness of the educational potential of technology and a willingness to apply their newly developed digital competence:

   S1 - "The vast knowledge of possible websites and applications"

   S2 - "The amount of websites that were shared and taught how to use"

   S4 - "I really enjoyed learning more about the many digital tools that exist"

   S6 - "Knowing tools that we can use in our classroom"

   S9 - "Learning about different platforms to use in the classroom, most of which I didn't know"

b) Pre-service teachers also acknowledged the potential of digital tools to improve student engagement and involvement:

   S2 - " (digital tools) allowed me to turn somewhat boring themes into interesting classes that still let students learn."

   S4 - "I believe these tools can be great to make the students more interested in participating and make them feel more comfortable in the classroom. These tools are intuitive for both the teacher and the students and I believe that is plus to why we should use them."

S5 - "(digital tools) capture the students' attention better, rather than the traditional method of viewing the whole video and only then answering questions normally on paper."

S6 - "(knowing digital tools) can encourage students to learn and provide a better and funnier learning experience."

c) Pre-service teachers expressed their satisfaction for having learnt to use digital technologies to create teaching and learning resources, emphasized their educational usefulness:

S2 - "by far the best source of practical knowledge I've received."

S3 - "It is beneficial and exciting"

S5 - "I also found it interesting"

S8 - "It was really interesting and even fun"

S11 - "most of the aspects of the module are quite beneficial"

d) One pre-service teacher expressed their gratitude for the module's emphasis on responsible use of digital tools and digital learning environments:

S10 - "I appreciate that we analyzed both the pros and cons of using digital tools for learning, highlighting that it only helps if we use it critically. "

"most of the aspects of the module are quite beneficial."

e) Participants underlined the benefits of the supportive learning environment and the adaptive teaching technique which enhanced the module's efficacy:

S8 - "The quality of the interaction between student and teacher was really a major factor on why I felt comfortable enough to share my ideas or just ask questions."

S12 - "The gradual project was quite useful for learning how to work with different tools at different times and at our own pace."

3.2   Question 2

a) Participants reiterated their appreciation of the course module and pointing out that it had given them the opportunity to develop their digital literacy skills:

S2 - "There isn't much to improve based on the nature of the module, it's a fresh and exciting experience though still not very tangible to us. This goes to say that being a future teacher with no previous experience turns the classes from this module into a fantasy play that I get to discover this year."

S3 - "The Digital Competence Module was an amazing idea! We really needed it. In my opinion everything was well prepared and the module helped me to learn and also apply the knowledge I accquired."

S4 - "I do not see any aspects that need to be improved, as I enjoyed this Module"

S11 - "Other than that, I believe that the Digital Competence Module was quite fruitful and productive."

b) Students then pointed out aspects of the course module which they would like to improve, such as getting more practice to develop a more practical use of digital technologies in the classroom:

S1 - "I think perhaps having more tasks to do at home and having tasks with more concrete goals"

S6 - "Maybe we could have given our activities for our colleagues to try and provide us with feedback."

S12 - "These lessons could have benefitted from more time to fully explore the extent of the available tools and to fully work on these digital projects with a lot more detail."

c) Finally, number of specific module improvements were proposed, such as a closer link between digital tools and subject content; attention to ethical issues; emphasis on lesser known digital resources and more time for research assignments:

S3 - "One thing that can be done to improve it is: try make more connections between the English Language and Culture Module"

S5 - "I think we should be more aware of copyright and digital licenses, as we can often inadvertently appropriate copyrighted material."

S8 - "I would have liked to research about the use of technologies on a classroom, perhaps picking a topic and writing an essay about it"

S9 - "time could've been used to learn about other websites."

S11 - "I believe that using Excel as a tool for evaluation and organisation would be quite beneficial to up-coming teachers"

4 Discussion and Implications

In this section, we will concentrate on the key findings and implications that form the core of our study.

Pre-service teachers acknowledged the potential of digital tools to engage and motivate learning in schools. This recognition signals a fundamental shift in the role of technology as triggering motivation and engagement in the learning process. Expanding on this, it becomes evident that ensuring a wide choice of digital tools is not merely a matter of technological diversity but a strategic response to the varied needs and preferences of today's learners [13]. This implies a departure from a one-size-fits-all approach to education, recognizing that students possess diverse learning styles and respond differently to various forms of digital content and that technology should be an enabler rather than a mere supplement to traditional teaching methods [14].

Pre-service teachers also acknowledged the module's practical approach, which enabled them to explore and experiment digital tools. This emphasis on experiential learning aligns with contemporary educational theories that highlight the significance of learning by doing [15]. It implies a departure from traditional instructional methods toward dynamic and interactive learning experiences. In the context of digital tools, this means providing opportunities for pre-service teachers to experiment, make mistakes, and refine their approaches based on practical insights gained from the application of these tools in simulated or actual classroom settings. Prioritizing experiential learning is, therefore, not just a suggestion but a transformative strategy in teacher education [16]. It is a call to action to reshape teacher training paradigms and ensure that the next generation of educators is not only well-informed but also proficient practitioners in the dynamic realm of digital teaching tools.

The participants' distinct emphasis on the critical thinking aspect and the importance of evaluate the advantages and disadvantages of digital tools signals a collective desire among educators to move beyond mere adoption of technology and, instead, to make well-informed decisions that align with pedagogical best practices [17, 18]. The call for critical thinking in the evaluation of digital tools also implies an ethical approach to technology integration in education, which clearly show that teachers are not seeking a one-size-fits-all solution but choose digital tools within the context of their specific pedagogical context. This signifies a departure from a technology-centric approach to a more thoughtful and strategic use [19, 20]. An important implication of this finding is that future editions of the course module should continue to encourage this evaluative mindset and enable teachers to critically assess the relevance, impact, and suitability of various tools for their specific teaching contexts. This also implies that future training should not only focus on popular or widely used tools but tools that are aligned with the learning goals.

5 Limitations

It is essential to recognize the limits of the study prior to making more generalizations. A significant constraint is the very small sample size, which limits the applicability of the results to a larger population. While it is crucial to acknowledge the inherent limitations of this study, it is important to contextualize these constraints within the study's goal, which was to gather insights into pre-service teachers' perceptions of a specific course module.

6 Conclusion

The goal of this study was to examine the strengths and potential weaknesses of a digital competence module developed for pre-service English language teachers. Feedback from the pre-service teachers highlighted positive aspects of the module, such as its effectiveness in introducing new tools and providing a hands-on application of such tools (Sect. 3.1). Pre-service also identified the need for more practical application and for attention to copyrights and other ethical challenges (Sect. 3.2). Within this context, the study provides valuable information that contributes meaningfully to our understanding of their needs and expectations. However, as mentioned above, the study's findings are based on a relatively small sample size, which necessarily limits the generalizability of the results. To address this limitation and enhance the robustness of our conclusions, future research will involve a larger and more diverse cohort of participants. By increasing the sample size, we aim to conduct a more comprehensive analysis that can yield statistically significant insights into the impact and effectiveness of the digital competence module.

References

1. Kirschner, P.A., De Bruyckere, P.: The myths of the digital native and the multitasker. Teach. Teach. Educ. **67**, 135–142 (2017)
2. Pettersson, F.: On the issues of digital competence in educational contexts–a review of literature. Educ. Inf. Technol. **23**(3), 1005–1021 (2018)

3. Mishra, P., Koehler, M.J.: Technological pedagogical content knowledge: a framework for teacher knowledge. Teach. Coll. Rec. **108**(6), 1017–1054 (2006)
4. Prensky, M.: Digital natives, digital immigrants part 1. Horizon **9**(5), 1–6 (2001)
5. Instefjord, E., Munthe, E.: Educating digitally competent teachers: a study of integration of professional digital competence in teacher education. Teach. Teach. Educ. **67**, 37–45 (2017)
6. OECD: Trends Shaping Education 2022, 2nd edn. OECD Publishing, Paris (2022)
7. Voogt, J., McKenney, S.: TPACK in teacher education: are we preparing teachers to use technology for early literacy? Technol. Pedagogy Educ. **26**(1), 69–83 (2017)
8. UNESCO: UNESCO ICT Competency Framework for Teachers, 2nd edn. UNESCO, Paris (2018)
9. Redecker, C.: European framework for the digital competence of educators: DigCompEdu. In: Punie, Y. (ed.) Publications Office of the European Union, Luxembourg (2017)
10. INTEF (The National Institute of Educational Technologies and Teacher Training): Common Digital Competence Framework for Teachers. https://aprende.intef.es/sites/default/files/2018-05/2017_1024-Common-Digital-CompetenceFramework-For-Teachers.pdf. Accessed 22 June 2022
11. Koehler, M.J., Mishra, P.: What is technological pedagogical content knowledge. Contemp. Issues Technol. Teacher Educ. **9**, 60–70 (2009)
12. European Commission/EACEA/Eurydice: Digital Education at School in Europe. Eurydice Report. Publications Office of the European Union, Luxembourg (2019)
13. Tondeur, J., van Braak, J., Sang, G., Voogt, J., Fisser, P., Ottenbreit-Leftwich, A.: Preparing pre-service teachers to integrate technology in education: a synthesis of qualitative evidence. Comput. Educ. **59**(1), 134–144 (2012)
14. Reisoglu, I.: How does digital competence training affect teachers' professional development and activities? Technol. Knowl. Learn. **27**, 721–748 (2021)
15. Brown, C.: Pedagogy and the New Literacies in Higher Education, pp. 792–805 (2014)
16. Laurillard, D.: The pedagogical challenges to collaborative technologies. Int. J. Comput.-Support. Collab. Learn. **4**, 5–20 (2009)
17. Leming, T., Johanson, L.B.: "And then I check to see if it looks legit" – digital critical competence in teacher education. Front. Educ. **8** (2023)
18. Coleman, H., Dickerson, J., Dotterer, D.: Critical thinking, instruction, and professional development for schools in the digital age, pp. 1235–1254 (2017)
19. Masoumi, D., Noroozi, O.: Developing early career teachers' professional digital competence: a systematic literature review. Eur. J. Teacher Educ., 1–23 (2023)
20. Weidlich, J., Kalz, M.: How well does teacher education prepare for teaching with technology? A TPACK-based investigation at a university of education. Eur. J. Teacher Educ., 1–21 (2023)

Empowering the Teaching and Learning of Geometry in Basic Education by Combining Extended Reality and Machine Learning

Carlos R. Cunha[1]([✉]) [iD], André Moreira[2] [iD], Sílvia Coelho[3] [iD], Vítor Mendonça[2] [iD], and João Pedro Gomes[2] [iD]

[1] Research Centre in Digitalization and Intelligent Robotics (CeDRI), Instituto Politécnico de Bragança, Campus de Santa Apolónia, 5300-253 Bragança, Portugal
crc@ipb.pt

[2] Instituto Politécnico de Bragança, Campus de Santa Apolónia, 5300-253 Bragança, Portugal
{andre-moreira,mendonca,jpgomes}@ipb.pt

[3] Escola Básica 2,3 Dr. Francisco Sanches, Rua do Taxa, 4710-448 Braga, Portugal
silvia.rc.coelho@sapo.pt

Abstract. Technology has helped to innovate in the teaching-learning process. Today's students are more demanding actors when it comes to the environment, they have at their disposal to learn, experiment and develop critical thinking. The area of mathematics has successively suffered from students' learning difficulties, whether due to lack of motivation, low abstraction ability or lack of new tools for teachers to bring innovation into the classroom and outside it. While being true that digitalization has entered schools, it often follows a process of digital replication of approaches and materials that were previously only available on physical media. This work focuses on the use of Extended Realities for teaching mathematics, and very particularly in the teaching of geometry, with a proposition of a conceptual model that combines the use of Extended Reality and Machine Learning. The proposed model was subject to prototyping, which is presented as a form of laboratory validation as a contribution to innovate the way of how the geometry teaching-learning process is developed, as well through the ability to obtain useful insights for teachers, and students, throughout the process.

Keywords: Teaching · Learning · Geometry · Extended Reality · Machine Learning

1 Introduction

Ensuring access to technology is a key step for schools to convert to digital. However, for this conversion process to be successful, it is vital that the focus is not on technology but on understanding how the technology can enable teaching and learning in an effective and inclusive way [1, 2].

Nowadays, educational projects include the physical manual as well as digital materials, such as the manual in digital format, question banks, worksheets and exams.

© The Author(s), under exclusive license to Springer Nature Switzerland AG 2024
Á. Rocha et al. (Eds.): WorldCIST 2024, LNNS 988, pp. 98–109, 2024.
https://doi.org/10.1007/978-3-031-60224-5_11

It would be expected the teaching and learning of Mathematics, and in particular Geometry, to be greatly strengthened which consequently would result in student's improved performance. However, in most cases this barely happens, often because, the so-called technological support materials used are just a "blunt" digitization of what already exists in paper format – and solving exercises by only "writing on the computer", using the same paradigms, stops being interesting after a while.

A different, but equally strong, obstacle shared by many mathematics teachers is the lack of time to work/deepen each topic present in the curricular plan, including Geometry. At the same time, they are faced with a lack of interest on the part of students, which result in low attention and concentration during classes, as well in the subjects covered by the teacher, in the tasks to be carried out and, even more worrisome, the lack of continuity of the studies at home. According to [3], the lack of adequate educational methods is becoming increasingly evident, leading to a decrease in students' interest in learning and critical thinking.

Without resolving these basic problems, there is no prospect of improvements in terms of knowledge acquisition and application. For instance, many students do not read the statements properly and adopt a "give up" mentality simply because a geometric figure is in a position strange to them, which is a frequent scenario in Geometry exams surrounding primitives. Very often they are unable to apply the Pythagorean theorem in space, as they reveal difficulties at the level of abstraction. A study on the difficulties of applying the Pythagorean theorem can be found in [4].

In the current days it is required a tripartite commitment between School, Parents and Students to boost the teaching-learning process. If one fails, the entire teaching/learning process is compromised. According to [5], ensuring communication and supporting tools among the different actors present in the school ecosystem is an essential issue to approach, and the use of appropriate technologies can bring these different actors closer.

Generally, Students become more interested when present with the application of Mathematics in real-life scenarios and, more importantly, in the world of work. They need to carry out project works with a methodology that excites them – using the right tools can give a strong boost to this entire process.

Technology plays a key role in shaping the future of education [6]. New technologies have undergone significant advances, making it possible today to integrate them into the teaching-learning process in a more valid way for educational practices in teaching. There is a wide variety of innovative equipment, programs and serious games that offer special meaning in the construction of knowledge while presenting themselves as appealing. When well chosen, used appropriately and sparingly, planned according to the characteristics/profile of the targeted students, and taking the necessary time, they are certainly a strong plus – motivating the student to learn, without compromising the basic principles of each subject.

Dynamic Geometry environments can promote the teaching and learning of Geometry. It can help overcome difficulties or even avoid some of them. In fact, according to [7], the use of Dynamic Geometry Systems (DGS) has attracted special attention within mathematics education. This way of approaching the study of Geometry allows students to carry out tasks with a more exploratory and investigative approach, creating an environment where those involved can interact, discuss ideas, formulate conjectures, solve problems and, consequently, obtain generalizations with relative ease while being able to justify the obtained results. When they become protagonists of their own learning, students develop characteristics that contribute to aspects of their social, personal and professional lives.

One of the advantages of using new technologies in teaching Geometry is having dynamic access to complex figures that would otherwise be difficult to view from another perspective – knowledge acquisition becomes quick, easy, interactive, and accompanied by logical reasoning. In a study carried out by [8], the results indicated that the use of new technologies such as Virtual Reality (VR) and Augmented Reality (AR) improve interactivity and students' interest in teaching/learning mathematics, contributing to a more efficient understanding of various concepts, such as the study of primitives, when compared to the use of traditional teaching methods.

Empowering teachers and students, through the use of appropriate technologies, is vital to improve the teaching/learning paradigm. Since mathematics is often perceived as something abstract, being hard to visualize its laws and dynamics, as well as its applicability, it becomes important to provide mechanisms which enables a more "physical" visualization of these same laws and applications.

However, it is equally important to the teacher, when using technology, to understand the way in which students conduct their learning process, being able to personalize their experience and drawing useful insights to better respond the needs of each student. In this way, teaching actors will be able to maintain high levels of motivation, as well to promote autonomous learning and exploration in the students, while using an immersive environment that enable critical thinking.

According to [9], immersive media presents benefits in terms of increasing motivation and expanding traditional teaching practices, involving students in different ways.

This article proposes a model based on the use of Extended Reality (XR), to support the immersiveness and objectification of the not always objective dynamics of mathematics and Machine Learning (ML) to personalize the student experience and generate knowledge that will help the teacher to better understand and map the students' difficulties. The potential of ML is studied by [10], where the role that this technology can have in the ability to predict students' performance is highlighted.

The proposed approach – a conceptual model that combines XR and ML – is a contribution to supporting an immersive vision based on computational intelligence, of the teaching-learning process. It is also presented a prototype of the model, presented for the teaching of mathematics, particularly Geometry, taught in the context of basic education.

2   Literature Background on the Use of Extended Reality and Machine Learning on Education

The use of technology to innovate in the teaching-learning process is nothing new. In fact, according to [11], integrating information technologies as a tool capable of assisting in education is mandatory, especially in an era in which there is a notable transition in the teaching-learning process.

In a study carried out by [12], which analyzes the role of reality-enhancing technologies for the teaching and learning of mathematics, some conclusions point to the need of promote the use of these technologies by both students and teachers while the later needs to know more than just about the tools used, but also how they can teach through it. Likewise, in a study carried out by [13], which analyzes, using a systematic review, the different approaches to learning with Immersive Virtual Reality (IVR), educational researchers are urged to consider existing approaches and rethink the process used to design IVR-based learning tasks to achieve their pedagogical goals. Being IVR, according to [14], a potential approach to transform traditional classrooms into immersive virtual reality scenarios that are effective in the learning process and for research purposes.

With regard to geometry teaching, the use of AR is described by several authors. According to [15], over the past decade, the use of AR has proliferated in the education sector. However, the number of articles that systematically reviewed the research trends in the implementation of AR for learning mathematics was reduced.

In [16], a structure of a learning environment system based on AR, supported by mobile devices, is presented. This system allows students to get assistance as well as tips to solve problems in the context of geometry. While in [17], is proposed a design, implementation and evaluation of an AR-based geometry learning application where interactions are based on hand gestures. However, the model does not include the use of ML to understand students' difficulties and/or personalize teaching-learning experiences.

Also, in [18], AR was used to develop material as a support mechanism for teaching and learning mathematics and examining its effectiveness. And, while noted a improve in the students' learning outcome, it is stated that the development of AR material presents some difficulties, being necessary to solve technical problems, proceed with the improvement of certain features, and be able to provide more clear instructions to users. Similarly with the previous ones, in this work, the use of AR and ML was also not combined.

While applied in primary school, in [19], is presented a work where AR was used to assist learning achievement, generate motivation and creativity for children. The results of this investigation showed a positive impact on the degree of students' satisfaction in their teaching-learning process, an increase in students' motivation to learn less popular subjects (e.g., such as geometry), while also being noticed an improve in their creative thinking.

In a study carried out by [20], which analyzed the potential of AR for Teaching Mathematics, more specifically for teaching Vector Geometry, the use of AR in mathematics classes was considered beneficial and fun by the teacher and students. The use of AR supported the development of spatial imagination, switch that can be difficult to evolve when using only traditional 2D material.

Although there are several studies and applications of XR, namely AR and VR, in the field of teaching, the approaches used do not combine the potential of XR with ML, reducing the ability to generate intelligence each time a student uses the application during their learning process. In this way, is not taken advantage of the full potential return that would come from knowing the study patterns, failures and approaches that students have when solving or experimenting with the 3D educational materials. The combination of ML with XR will allow students to better understand their errors while, for teachers, it will provide a personalized and manageable teaching tool, with a positive impact on the readjustment of methodologies and pedagogy used in the classroom, combined with innovation.

ML can help understand the student and support the provision of adapted and personalized content, materializing an efficient recommendation system in the learning environment, while considering student's behaviors and preferences when recommending learning materials - a system that adapts to the needs and student's learning skills [21].

3   Proposed Conceptual Model

This section presents a conceptual model (see Fig. 1) based on the use of XR with ML and intended to be a contribution in innovating the teaching-learning process. The current model is applicable in any area of teaching; however, the developed prototype was targeted to the teaching of mathematics and more specifically for the teaching of geometry in Portuguese basic education.

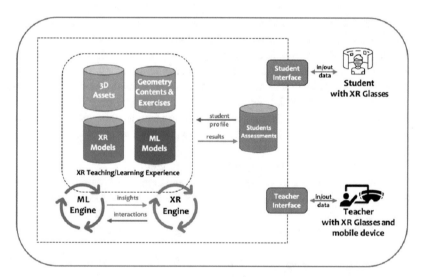

Fig. 1. Proposed Conceptual Model.

The proposed model has several components and actors which are presented below, and where their main objectives and form of action are characterized:

3D Assets Database: Represents the repository of 3D assets that are used in XR applications (e.g., solid primitives).

Geometry Contents & Exercises: It represents the content associated with the subject that will be taught and/or the object of study to be developed inside or outside the classroom. This content will make use of the 3D assets to support XR applications.

XR Models: It represents the set of scenarios that will be personalized by the teacher's depending on the context of the classroom, homework or free study activities to be carried out by students.

ML Models: It represents a group of models that has been trained to recognize certain types of patterns obtained through data sets and algorithms that are used to weight and learn from stored data (which are obtained from the data trail that is left by students when using XR applications and keep on a interoperable database)

XR Engine: Represents the module that extracts the actions carried out by student and learner (being a teacher or a colleague), feeding the ML engine. Being capable of generating personalized XR scenarios based on input from the ML engine.

ML Engine: Represents the intelligence engine of the proposed model. It is capable of, analyze the insights and actions carried out by students when carrying out activities, and interact with the XR Model to generate new scenarios with personalized activities, based on the difficulties of each student, as well as, in real time, the generation of each new step of the activity (e.g., next question).

Students Assessments: Represents the repository of assessments and results of these from each student with the finality of assisting the teacher in understanding students' difficulties.

Student Interface: Represents the students-view of the system. Students will use XR Glasses (e.g., HoloLens® 2), and all available functionalities for them to use (e.g., collaborative working in classroom, homework, free activities at home).

Teacher Interface: Represents the teacher-view of the system. Teachers will use XR Glasses (e.g., HoloLens® 2), being able to use the same functionalities of the students as well as the ones directed to them on an administrative/manager level.

In order to exemplify some functionalities to be implemented, the following use case diagram (see Fig. 2) illustrates the activities involved in classroom learning.

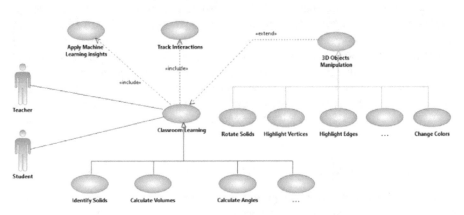

Fig. 2. Use Case – Classroom Learning.

Thus, as illustrated, it can be seen that the system constantly tracks the interactions that will serve as the basis for machine learning insights. In classroom learning, both the student and the teacher will be able to interact, allowing different types of learning (identification of solids, areas and volumes calculation, among others) while being able to manipulate objects in different ways (rotate solids, highlight vertices and/or edges, change colors, among others).

On the other hand, the system should allow the teacher to monitor the progress of students' learning, as represented in the following use case diagram (see Fig. 3).

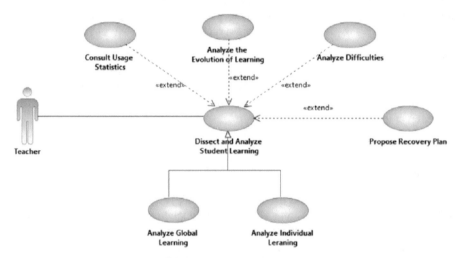

Fig. 3. Use Case – Analyze the evolution of learning.

Monitoring the evolution of learning, using statistical analysis and inference which will result in a better understanding of the difficulties experienced by students, will also allow the teacher to propose new challenges and recovery plans.

The conceptual model presented is a contribution to support innovation, we idealize and operationalize new teaching-learning paradigms, more capable of meeting students' motivations and in contributing to their critical development. Likewise, it aims to contribute to the creation of collaborative environments in the classroom and to help break barriers in the difficulty of abstraction associated with mathematics and especially in the teaching of Geometry, where, most times, the 2D view of materials on paper stands as an obstacle instead of support. Finally, it aims to help generate motivation in students to develop independent study work while providing valuable insights for the teacher at an interoperable level.

4   Prototype Developed

This section presents the developed prototype based on the conceptual model presented in the previous section being referred the technologies used, the implemented features and planned ones as well. This is an ongoing work that aims the validation of the conceptual model proposed, and to be tested in real school classes context.

4.1 Technologies Used

When developing the prototype was considered the available equipment for Mixed Reality (MR) being this the HoloLens® 2, an Optical Head-Mounted Device (HMD) or Optical See-Through (OST) HMD [22], manufactured by Microsoft®, this device was released in 2019 and it works as a standalone HMD without the need of controllers or connection to external components, being this possible by featuring technologies like Hand Tracking, Eye Tracking and Spatial Mapping.

Unity. Also known as Unity3D or Unity Engine, is a Game Engine used primarily for game development, being highly recognized among the XR Developers Community duo to his low learning curve and ease-of-use when compared with others game engines like Unreal or CryEngine, the big community and support when developing XR applications as well for being a primary suggestion by most of the pioneers of this technology like Meta®, Microsoft® and VIVE®.

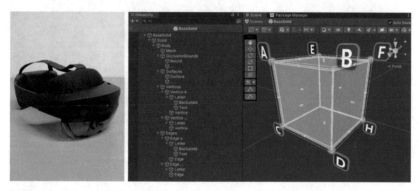

Fig. 4. HoloLens® 2 and Prototype 3D Primitive GameObject Architecture with Example in Unity Editor.

MRTK3. The Mixed Reality Toolkit is a framework created by Microsoft® to provide the necessary tools to developers when creating MR applications to HoloLens® 2 in Unity, the current version is the MRTK3 and while available as public preview since June 2022, it was declared as generally available in September 2023. While focused on the HoloLens® 2, applications developed with MRTK2 and principally MRTK3 can also be built to be used in other devices that support MR like Magic Leap 2, Meta® Quest 2 and the most recent Meta® Quest 3.

4.2 Features Implemented

In this section will be described the implementation process of the three main concepts presents in the developed prototype.

General Solid Architecture. Geometry is the branch of mathematics concerned with the study of solids, lines, faces and vertices, as such these elements are the very base necessity to be replicated in MR being also necessary the possibility of treating each component as individual element. As such the solid, being a cube or a triangular prism for example, is mounted by defining each component individually (see Fig. 4), the vertices are defined by using 3D Vectors (Vector3) while the edges make use of a feature of Unity called Line Renderer being used the vertices to define the start and end of the line. However, for the definition of faces, as well to have the complete solid as individual component for some use cases and for more complex's solids, was necessary the use of primitives that Unity does not provides, being acquired the asset Deluxe Primitives by Reactorcore Games and Ultimate Procedural Primitives by KANIYONIKA from the Unity Asset Store.

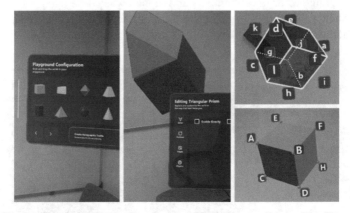

Fig. 5. 3D Primitives Creation screen and Solid Personalization Use-Cases

Solid Personalization. Consists in empowering the user, being a teacher or student, by enabling him to modify the solid in a way to better highlight a concept (see Fig. 5).

Change Color. Enables the user to highlight specific components of the solid or even differentiate them being a basic concept present both in school manuals or when used physical solid representations, but without limitations or constraints to presets.

Edit Vertices. Allows the user to modify the representation of the vertices, while they are defined through simple 3D points, the user can choose to display visual markers (spheres), or letters as well display only specific ones.

Edit Edges. Provides a group of possibilities such as display only the edges while hiding the solid faces as well display letters. The user can also choose to display only the outline of the solid or enable invisible edges.

Physics. Allows changing solid characteristics such as how it interacts with gravity, and mass, as well with other solids, which can be used for explaining geometric concepts such as cube number.

Playground Configuration. Refers to the basic ability of the user to create more solids at his own choice and to manipulate them both in terms of position, scale, and rotation in the digital world, as well to create composite solids. Depending on how the solid physics is configured those same solids can collide between them or be inserted inside each other to assist for example at calculating volumes.

4.3 Ongoing and Future Development

Aside from further improvements, based on the necessities noted during the initial tests, like the possibility of personalizing the surfaces to for example only being visible a set of them, the possibility of creating 2D primitives such as triangles instead of only 3D or clipping 3D primitives.

Some more advanced features planned are the possibility of streaming the user view to a desktop screen using Mixed Reality Capture (MRC) and WebRTC, enabling scenarios where the teacher may present concepts during classes while projecting the 3D solids for the class. Another feature is the development of shared experiences in a way that two or more users of the MR HMD may interact with the same 3D world similarly to Multiplayer Games by using technologies like Photon Unity Networking 2 (PUN 2).

Finally, the ML engine is being implemented and will be improved after experimentation in a real scenario (i.e., school classes) and, being defined objectives, in partnership with basic school teachers, considered most relevant to help personalize the proposed exercises and which will help to understand the reasons that tend to characterize the difficulties in understanding and learning, of each student.

5 Conclusion

Technology has supported and encouraged the introduction of innovation in education, promoting new methodologies inside and outside the classroom, as well as innovating the concepts of school manuals and study support material.

Among the technologies that are most promising, XR can enhance essential factors in the teaching-learning process, such as increasing levels of motivation, capacity of abstraction and development of critical thinking.

Likewise, the use of XR to create cooperative teaching-learning environments allows the construction of a new paradigm of interaction between teacher and students, especially in the classroom.

The combination of XR with ML raises the potential of XR, allowing the acquisition of insights that can be decisive in understanding, on the one hand, students' difficulties in a personalized way and, on the other, the definition of more appropriate strategies inside and outside the classroom.

This paper proposes a conceptual model based on the use of XR and ML, as a contribution to the creation of new and innovative teaching-learning strategies. In order to validate the presented model, a prototype was developed based on MR to assist the teaching Geometry. Being still under development for full validation of the model. Relatively to the ML component, it intends to be implemented after a period of experimentation in a classroom with subsequent validation of the MR approach and after being received and analyzed the feedback from the involved actors.

Acknowledgements. The authors are grateful to the Foundation for Science and Technology (FCT, Portugal) for financial support through national funds FCT/MCTES (PIDDAC) to CeDRI (UIDB/05757/2020 and UIDP/05757/2020) and SusTEC (LA/P/0007/2021).

References

1. McKnight, K., O'Malley, K., Ruzic, R., Horsley, M.K., Franey, J.J., Bassett, K.: Teaching in a digital age: How educators use technology to improve student learning. J. Res. Technol. Educ. **48**(3), 194–211 (2016)
2. Dahal, N., Manandhar, N.K., Luitel, L., Luitel, B.C., Pant, B.P., Shrestha, I.M.: ICT tools for remote teaching and learning mathematics: a proposal for autonomy and engagements. Adv. Mobile Learn. Educ. Res. **2**(1), 289–296 (2022)
3. Georgieva, L., Nikulin, A.: The art of education: Creative thinking and video games. Balkan J. Philos. **15**(2), 179–186 (2023)
4. Sari, R.H.Y., Wutsqa, D.U.: Analysis of student's error in resolving the Pythagoras problems. J. Phys. Conf. Ser. **1320**(1), 012056 (2019). https://doi.org/10.1088/1742-6596/1320/1/012056
5. Cunha, C.R., Gomes, J.P., Mendonça, V.: Bringing together high school actors using mobile applications: a conceptual model proposal. In: Proceedings of the 34th International Business Information Management Association Conference, International Business Information Management Association, pp. 11564–11570 (2019)
6. Haleem, A., Javaid, M., Qadri, M.A., Suman, R.: Understanding the role of digital technologies in education: a review. Sustainable Operations Comput. **3**, 275–285 (2022). https://doi.org/10.1016/j.susoc.2022.05.004
7. Bozkurt, G., Uygan, C.: Lesson hiccups during the development of teaching schemes: a novice technology-using mathematics teacher's professional instrumental genesis of dynamic geometry. ZDM, 1–15 (2020)
8. Demitriadou, E., Stavroulia, K.-E., Lanitis, A.: Comparative evaluation of virtual and augmented reality for teaching mathematics in primary education. Educ. Inf. Technol. (Dordr.) **25**, 381–401 (2020)
9. Erturk, E., Reynolds, G.-B.: The expanding role of immersive media in education. In: International Conference on E-learning, pp. 191–194 (2020)
10. Albreiki, B., Zaki, N., Alashwal, H.: A systematic literature review of student'performance prediction using machine learning techniques. Educ. Sci. (Basel) **11**(9), 552 (2021)
11. Laseinde, O.T., Dada, D.: Enhancing teaching and learning in STEM Labs: the development of an android-based virtual reality platform. Mater Today Proc. (2023). https://doi.org/10.1016/j.matpr.2023.09.020
12. Buentello-Montoya, D.A., Lomelí-Plascencia, M.G., Medina-Herrera, L.M.: The role of reality enhancing technologies in teaching and learning of mathematics. Comput. Electr. Eng. **94**, 107287 (2021)
13. Won, M., et al.: Diverse approaches to learning with immersive virtual reality identified from a systematic review. Comput. Educ. **195**, 104701 (2023)
14. Hasenbein, L., et al.: Learning with simulated virtual classmates: effects of social-related configurations on students' visual attention and learning experiences in an immersive virtual reality classroom. Comput Human Behav. **133**, 107282 (2022). https://doi.org/10.1016/j.chb.2022.107282
15. Ahmad, N., Junaini, S.: Augmented reality for learning mathematics: a systematic literature review. Inter. J. Emerging Technol. Learn. (iJET) **15**(16), 106–122 (2020)

16. Gargrish, S., Mantri, A., Kaur, D.P.: Augmented reality-based learning environment to enhance teaching-learning experience in geometry education. Proc. Comput. Sci. **172**, 1039–1046 (2020). https://doi.org/10.1016/j.procs.2020.05.152
17. Le, H.-Q., Kim, J.-I.: "An augmented reality application with hand gestures for learning 3D geometry. In: 2017 IEEE International Conference on Big Data and Smart Computing (BigComp), pp. 34–41 (2017). https://doi.org/10.1109/BIGCOMP.2017.7881712
18. Koparan, T., Dinar, H., Koparan, E.T., Haldan, Z.S.: Integrating augmented reality into mathematics teaching and learning and examining its effectiveness. Think Skills Creat **47**, 101245 (2023). https://doi.org/10.1016/j.tsc.2023.101245
19. Yousef, A.M.F.: Augmented reality assisted learning achievement, motivation, and creativity for children of low-grade in primary school. J. Comput. Assist. Learn. **37**(4), 966–977 (2021)
20. Schutera, S., et al.: On the potential of augmented reality for mathematics teaching with the application cleARmaths. Educ. Sci. (Basel) **11**(8), 368 (2021)
21. Oubalahcen, H., Tamym, L., lay Driss El Ouadghiri, M.: The use of AI in E-Learning recommender systems: a comprehensive survey. Proc. Comput. Sci. **224**, 437–442, (2023). https://doi.org/10.1016/j.procs.2023.09.061
22. Doughty, M., Ghugre, N.R., Wright, G.A.: Augmenting performance: a systematic review of optical see-through head-mounted displays in surgery. J Imaging **8**(7) (2022). https://doi.org/10.3390/jimaging8070203

(Re)thinking Teacher Training in the Digital Age: Teacher Training Models for Online Practice

Bruno F. Gonçalves[1]([⊠]) [iD], Piedade Vaz-Rebelo[2] [iD], Maria Teresa R. Pessôa[2] [iD], and Vítor Gonçalves[1] [iD]

[1] Basic Education Research Center, Polytechnic Institute of Bragança, Bragança, Portugal
bruno.goncalves@ipb.pt

[2] Center for Interdisciplinary Studies of the 20th Century, University of Coimbra, Coimbra, Portugal

Abstract. In recent years, the need to train teachers to teach online has been highlighted, not only because of the exponential growth of technologies in all economic sectors of society, but also because of the very need for education to modernize and, through this, innovate in the digital sphere. The covid-19 pandemic has only demonstrated, now unequivocally, the urgency of teacher training in the field of online teaching, as well as the need for instructional design and evaluation of training models to support learning. It is precisely because of these two assumptions that there is an urgent need to develop research in the area that allows us to (re)think teacher training in the digital age so that teachers can participate in training actions and, through them, acquire fundamental skills for the realization of online teaching. In this sense, through a systematic review of the literature, we sought to identify and characterize training models geared towards teaching practice in online education. In this way, it may be possible to contribute to the development of a matrix for the instructional design of a teacher training model that prepares these professionals for online education. The results show that, although the literature has made an indispensable contribution to reflection on teacher training models, we recognize that more research is needed in the area, since the literature is deficient in terms of the subject under analysis.

Keywords: Digital technologies · Instructional design · Online teaching · Teacher training · Training models

1 Introduction

In the last two decades in particular, several studies and references have appeared in the literature on the competences that teachers must have in order to exercise the teaching profession [1–6]. However, there is still very little research into the digital competences of teachers in online teaching or the most suitable models for both online teaching and teacher training for this practice. Thus, although the last few decades have shown a steady growth in online education, with institutions offering more online courses and

Á. Rocha et al. (Eds.): WorldCIST 2024, LNNS 988, pp. 110–119, 2024.
https://doi.org/10.1007/978-3-031-60224-5_12

programs [7, 8], there are still numerous questions about online teaching. In fact, we have good examples at international level of higher education institutions that have been very successful in implementing online teaching in some of the curricular units they offer, for example: Queen's University, Massachusetts Institute of Technology (MIT), National Taiwan University, Stanford University, University of Melbourne, Duke University, among others. At a national level, not many institutions offer training in online education, but the work done in this area by the Universidade Aberta (uab) is recognized. Perhaps because there is still little use of online learning, the Portuguese government has set itself targets of training around 3,000 people by 2023 and up to 50,000 people by 2030 in distance learning [9]. Based on the assumption that these objectives should actually be met, it is becoming urgent to address this issue, because if the number of students in online education increases, more teachers are needed to train these students. Although experts in the field have identified many techniques, methods and approaches to help train and support online instructors [10], they can now be improved to meet the objectives set by the Portuguese government [9] and the sustainable development goals identified by the United Nations [11].

In this sense, there is a need to develop a systematic literature review that will allow us to identify and characterize the models oriented towards teacher training for online teaching, but also to understand what has been done in the area so far. In fact, there are some recent studies [12–19] that have contributed to literacy in the area and to changing the current paradigm. In addition to the United Nations' goals [11] for sustainable development and the Portuguese government's goals [9] especially for 2030, other studies that have been developed also point to the urgency of training teachers for online teaching. These studies will be duly addressed in the results section.

2 Methodology

Through this research we sought to carry out a systematic literature review with the aim of identifying and characterizing training models for pedagogical practice in online teaching. For this, the systematic literature review was adopted as the investigative methodology to support the study. The systematic literature review will be carried out with the support of a pre-defined set of criteria that will be essential for the selection of information on the theme addressed, namely: (i) Time interval: 2018–2023; (ii) Documents: reference articles and doctoral thesis; (iii) Search language: Portuguese and English; (iv) Bibliometric databases: Scopus and Web of Science; (v) Other databases: Google Scholar; Scielo; Springer; B-on; (vi) Keywords in Portuguese: "Modelos de formação de professores para a prática online"; "Formação de professores para o ensino online"; "Desenvolvimento profissional de professores para a prática online"; "Modelos de formação de professores"; "Modelos de desenvolvimento professional de professores" (vii) Keywords in English: "Models of teacher training for online practice"; "Teacher training for online teaching"; "Professional development of teachers for online practice"; "Models of teacher training"; "Models of teacher professional development"; (viii) All teachers at all levels of education, with the exception of higher education.

It should be noted that we initially tried to find models of teacher training geared towards practice in online teaching, but we found that there is little literature on this subject. We therefore decided to extend the study and include two more keywords: "Models of teacher training"; "Models of teacher professional development".

Based on the criteria previously established, the general framework of the documents found is presented (Table 1):

Table 1. General table of documents found.

ID	Article name	Year	Authors
1	Concepções e modelos de formação de professores: reflexões e potencialidades	2018	Raul Sardinha Netto & Maria Antonia Ramos de Azevedo [20]
2	Teacher professional development models for effective teaching and learning in schools	2019	Ceren Çetin & Mustafa Bayrakcı [21]
3	A professional development process model for online and blended learning: Introducing digital capital	2019	Brent Philipsen [22]
4	Improving teacher professional development for online and blended learning: a systematic meta-aggregative review	2019	Brent Philipsen, Jo Tondeur, Natalie Pareja Roblin, Silke Vanslambrouck, Chang Zhu [17]
5	Os Modelos de Formação de Professores/as da Educação Básica: quem formamos?	2020	Camila Lima Coimbra [23]
6	A pedagogical model for effective online teacher professional development—findings from the Teacher Academy initiative of the European Commission	2020	Benjamin Hertz, Hannah Grainger Clemson, Daniella Tasic Hansen, Diana Laurillard, Madeleine Murray, Luis Fernandes, Anne Gilleran, Diego Rojas Ruiz, Danguole Rutkauskiene [24]
7	Professional Development for Online Teaching: A Literature Review	2020	Leary, H., Dopp, C., Turley, C., Cheney, M., Simmons, Z., Graham, C. R., & Hatch, R. [25]
8	Innovating teachers' professional learning through digital technologies	2020	Andreea Minea-Pic [26]
9	Successful design and delivery of online professional development for teachers: A systematic review of the literature	2021	Leicha A. Bragg, Chris Walsh, Marion Heyeres [27]
10	Modelos formativos da docência: considerações acerca das racionalidades técnica, prática e crítico- reflexiva na formação de professores	2021	Caio Corrêa Derossi, Karen Laíssa Marcílio Ferreira [28]

(*continued*)

Table 1. (*continued*)

ID	Article name	Year	Authors
11	Modelos de Formação Docente: movimentos e reflexões para uma abordagem teórico-prática	2021	Alexander Montero Cunha [29]
12	Blended Learning in Teacher Education & Training	2021	Eileen Kennedy [30]
13	Teaching in Secondary Education Teacher Training with a Hybrid Model: Students' Perceptions	2022	José Luis Martín-Núñez, Juan Luis Bravo-Ramos, Susana Sastre-Merino, Iciar Pablo-Lerchundi, Arturo Caravantes Redondo and Cristina Núñez-del-Río [31]
14	Distance Education for Teacher Training: Modes, Models, and Methods	2023	Mary Burns [32]
15	Online Learning Standards: Steps to Introduce a Distributed Leadership Approach to Training Teachers for Online Teaching and Learning	2023	Geraldine Grimes, Fiona Boyle, Michael Noctor [33]

Eight documents were excluded because they were outside the lines of the research and, in this sense, fifteen documents were considered for the purposes of this research.

The data collected from the documents that emerged from the search were categorized and treated in Microsoft Excel. The study of the documents considered for the purposes of the research was carried out through content analysis, focusing on the identification and characterization of the proposed teacher training models.

3   Teacher Training Models

A few documents were found that addressed some of the models of teacher training, but not many articles addressed the practice of teachers in online teaching.

Thirty-three years ago, [34] identified a set of models aimed at the professional development of teachers, namely: observation/supervision-based, autonomous, training courses, curriculum and organizational development and action research. As early as 1997, [35] highlighted a set of models oriented towards the professional development of teachers, namely: the model of socialization into the professional culture, the technical model, the teaching model, the model that emphasizes the relationship between the personal and professional domains in teachers' work and the reflective model or practical reflection.

In the 21st century, [36] presented another model, called the interconnected model of teacher professional growth, which is divided into four distinct domains that change through the processes of 'reflection' and 'implementation', namely: the personal domain, the domain of teaching practices, the consequences for student learning and the external domain [36].

A little later, [37] identified two groups of models for teacher professional development: the first group considers professional development contexts and strategies that make use of inter-institutional cooperation, such as university-school partnerships; the second group refers to professional development strategies that do not require institutional cooperation, such as supervision of teaching practices.

[38] also identify three more models, namely: standardized development, which focuses on specific skills and content and is aimed at training trainers; self-directed, which allows for individual learning and a non-formal structure; and intensive learning by groups of teachers (learning communities), based on activities and aimed at long-term change processes [38].

In 2007, [39] presented a model describing the process of teacher professional development, the strategies to be implemented in the teaching-learning process and the evaluation of the results. The contents of the model (the curriculum, assessment and teaching) are discussed in professional learning communities, through a process of practice, feedback, reflection and evaluation of results.

In 2009, [40] proposed a typology of four models, namely: re-instrumentation, remodeling, revitalization and re-imagination [40]. Re-instrumentation corresponds to the acquisition of skills and competencies for teaching, which are understood as predominantly technical activities; remodeling integrates programs that are normally associated with curricular reforms and other changes; revitalization corresponds to the individual learning of each teacher as well as their reflexivity; and, finally, the re-imagination foresees diversified learning paths, focusing on the political and activist dimension of the teaching profession [41].

In the same year, [42], in a study related to the professional development of teachers in ICT, refer to the existence of the following set of models: school-based provision, external provision, communities of practice, teacher inquiry model, critical reflection model and case-making model.

In a recent article [20], the authors address three interesting models: the technical model, the practical model and the critical model. The technician is based on an instrumental view of education and teacher training is aimed at mastering these techniques and methods. The practical emphasizes the importance of practical experience in teacher training. In this sense, training is aimed at developing practical skills and reflecting on one's own practice. The critic starts from a broader view of education, in which the teacher is seen as an agent of social transformation. Teacher training is focused on critical reflection on one's own practice and on the social relations that permeate education.

Another recent article [26] identifies three models of teacher training: online learning, community learning and blended learning professional training. Online learning involves the use of digital technologies to provide professional learning opportunities for teachers, such as MOOC [43–45]. Learning communities promote teacher participation in online communities or teacher networks for teacher training. Participation in teacher networks has been found to be a more effective form of learning than more traditional forms of teacher training [46, 47]. The blended learning professional training combines face-to-face and online learning elements. Combining in-person and online interactions can improve teachers' professional development [43, 44].

A group of authors [25] who discuss the professional development of teachers for online teaching highlight workshops, self-learning, peer mentoring, online modules and the creation of courses as suitable models for teacher training. It is also important to note that the article highlights the importance of understanding instructors' needs and best practices for online teaching, as well as the appropriate use of different training methods.

Another group of authors [24] suggests that collaborative approach to online contin-ued professional development can be developed as a way of addressing both teachers and education system needs. They highlight MOOC as a suitable model for teacher training.

[21] identify several models of professional development, namely: Individually guided development, observation and assessment, involvement in a development or improvement process, training, inquiry, mentoring, critical friends' group and profes-sional development schools. Each of these models has advantages and disadvantages and it is up to teachers to select the one that best suits their reality and needs.

A paper on blended-learning in teacher education and training [30] discusses the successful implementation of teacher professional development programs using blended learning approaches, as well as practical examples of their implementation in Europe. The article also presents additional information on teacher training models, such as the article by [48] on cooperative learning of teachers and teaching students and the article by [49] on teacher supply and demand issues in northern Canada.

The document "Distance Education For Teacher Training: Modes, Models and Meth-ods" [32] presents six distance education models that we think are quite interesting and organized, namely: print-based distance education, audio-based distance education, visually-based distance education, multimedia-based distance education, online learn-ing, mobile-based distance education. If we focus especially on the last two models, it appears that both already integrate digital technologies in the teacher training process.

It is important to highlight two models that are present in "Distance Education For Teacher Training: Modes, Models and Methods", namely online learning and mobile-based distance education. These models seem to be useful since they already take digital technologies into account in teacher training. Online learning presents itself as a scal-able model that can be easily updated, provides immediate feedback and contributes to reducing the isolation of new teachers. However, its limitations include the need for regular access to computers and the internet, as well as some quality control issues. This model can be seen as a cheap and easy solution. The mobile-based distance education model, on the other hand, has accessibility and portability as its main strengths. It is eas-ily distributed and can capitalize on technologies that teachers already have. However, it does have some associated limitations, such as the need for regular access to electricity and a cellular network or the Internet. It can also present problems regarding quality control and can easily be lost or stolen. However, these two models seem to be the ones that are currently attracting the most interest in teacher training.

Although each of the models presented identifies relevant aspects for the teacher training process, some gaps are recognized, especially regarding the integration of ICT in the teaching-learning process. In this sense, teachers must consider the benefits of each of the models or even combine them into several so that they include different opportunities and activities and, thus, it is possible to achieve the objectives of the professional and the educational institution. In other words, according to [29], we must take into account the importance of thinking about the detailed characteristics of training based on the context it has and how restrictive it can be to think about training by framing a priori the model you intend to use.

Many of the articles do not point to concrete models, but reflect on the importance of innovative practices in teacher training [50], instructional approaches, learning and assessment communities [27], the distributed leadership approach to teacher training [33], thinking and reflection on teacher training [29]. In addition to these, others stand out, such as [51] who address the effects of strategies in teacher training. The same author, in another article [17], presents a five-phase professional development process model for OBL, offering a new approach. Finally, [23] presents and discusses three models, namely: content model, transition model and resistance model.

4   Conclusions

In this research we seek to rethink teacher training in the digital age by identifying and characterizing teacher training models for online practice. By analyzing and discussing these models, we have been able to reflect on the main characteristics of each one. These training models aim to enable teachers to face the specific challenges associated with the virtual learning environment. More specifically, the aim is to enable teachers to improve their pedagogical practices, adapt to changes in the field of education and provide more effective learning experiences for students. Thus, by implementing these models, teachers can develop the specific skills needed to offer quality education in the online environment, considering the unique characteristics of this medium. It is important to note that an effective professional development approach often combines elements of several models, adapting to the specific needs of teachers and the dynamics of the educational institution. Combining several models can contribute to a greater diversity of pedagogical approaches, meeting individual needs, integrating diverse skills and greater flexibility and resilience, among other very useful advantages for teachers.

This research also allows us to conclude that the successful implementation of these professional development models for online practice depends on the ability to address the specificities of the virtual environment, promoting not only adaptation to technological tools, but also improving pedagogical strategies to create meaningful and effective online learning experiences. In this sense, it is also clear how important it is for teachers to actively engage in the learning that takes place in this type of environment, which aims to improve the skills of these professionals to improve the quality of the teaching-learning process in online environments.

Acknowledgment. This work has been supported by FCT – Fundação para a Ciência e Tecnologia within the Project Scope: UIDB/05777/2020.

References

1. Silva, E., Loureiro, M.J., Pischetola, M.: Competências digitais de professores do estado do Paraná (Brasil). EduSer **11**(1), 61–75 (2019)

2. UNESCO: Competency Standards Modules. ICT competency standards for teachers: competency standards modules. UNESCO, Paris (2009). http://unesdoc.unesco.org/images/0015/001562/156207por.pdf
3. Carretero, S., Vuorikari, R., Punie, Y.: DigComp 2.1: the digital competence framework for citizens with eight proficiency levels and examples of use. Joint Research Centre (Seville site) (2017)
4. Ainley, J., Carstens, R.: Teaching and learning international survey (TALIS) 2018 conceptual framework (2018)
5. UNESCO: UNESCO ICT Competency Framework for Teachers. Paris United Nations Education (2011)
6. Redecker, C.: European Framework for the Digital Competence of Educators: DigCompEdu. Publications Office of the European Union, Luxembourg (Luxembourg) (2017). https://doi.org/10.2760/178382, https://doi.org/10.2760/159770 (online)
7. Allen, I.E., Seaman, J.: Changing Course: Ten Years of Tracking Online Education in the United States. ERIC (2013)
8. Allen, I.E., Seaman, J.: Online Report Card: Tracking Online Education in the United States. ERIC (2016)
9. Governo de Portugal: Um 'contrato para a Legislatura' com o Ensino Superior para 2020–2023, orientado para estimular a convergência de Portugal com a Europa até 2030 (2019)
10. Lackey, K.: Faculty development: an analysis of current and effective training strategies for preparing faculty to teach online. Online J. Distance Learn. Adm. 14(4), 8 (2011)
11. United Nations General Assembly: Resolution adopted by the General Assembly on 25 September 2015. United Nations, Washington (2015)
12. Allela, M., Ogange, B., Junaid, M., Prince, B.: Evaluating the Effectiveness of a Multi-modal Approach to the Design and Integration of Microlearning Resources in In-Service Teacher Training (2019)
13. Brinkley-Etzkorn, K.E.: Learning to teach online: measuring the influence of faculty development training on teaching effectiveness through a TPACK lens. Internet High. Educ. 38, 28–35 (2018)
14. Brown, J., Brock, B., Závodská, A.: Higher Education in the 21st century: A New Paradigm of Teaching, Learning and Credit Acquisition. In: Proceedings of 14th IAC 2019, p. 87 (2019)
15. Gegenfurtner, A., Ebner, C.: Webinars in higher education and professional training: a meta-analysis and systematic review of randomized controlled trials. Educ. Res. Rev. 100293 (2019)
16. Kebritchi, M., Lipschuetz, A., Santiague, L.: Issues and challenges for teaching successful online courses in higher education: a literature review. J. Educ. Technol. Syst. 46(1), 4–29 (2017)
17. Philipsen, B., Tondeur, J., Roblin, N.P., Vanslambrouck, S., Zhu, C.: Improving teacher professional development for online and blended learning: a systematic meta-aggregative review. Educ. Technol. Res. Dev. 67(5), 1145–1174 (2019)
18. Singh, R.N., Hurley, D.: The effectiveness of teaching and learning process in online education as perceived by university faculty and instructional technology professionals. J. Teach. Learn. Technol. 6(1), 65–75 (2017)
19. Vagarinho, J.P., Llamas-Nistal, M.: Process-oriented quality in e-learning: a proposal for a global model. IEEE Access 8, 13710–13734 (2020)
20. Netto, R.S., de Azevedo, M.A.R.: Concepções e modelos de formação de professores: reflexões e potencialidades. Bol. Técnico do Senac 44(2) (2018)
21. Çetin, C., Bayrakcı, M.: Teacher professional development models for effective teaching and learning in schools. Online J. Qual. High. Educ. 6(1), 32–38 (2019)
22. Philipsen, B.: A professional development process model for online and blended learning: introducing digital capital. Contemp. Issues Technol. Teach. Educ. 19(4), 850–867 (2019)

23. Coimbra, C.L.: Os Modelos de Formação de Professores/as da Educação Básica: quem formamos? Educ. Real. **45**(1) (2020)
24. Hertz, B., et al.: A pedagogical model for effective online teacher professional development—findings from the Teacher Academy initiative of the European Commission. Eur. J. Educ. **57**(1), 142–159 (2022)
25. Leary, H., et al.: Professional development for online teaching: a literature review. Online Learn. **24**(4), 254–275 (2020)
26. Minea-Pic, A.: Innovating teachers' professional learning through digital technologies (2020)
27. Bragg, L.A., Walsh, C., Heyeres, M.: Successful design and delivery of online professional development for teachers: a systematic review of the literature. Comput. Educ. **166**, 104158 (2021)
28. Derossi, C.C., Ferreira, K.L.M.: MODELOS FORMATIVOS DA DOCÊNCIA: CONSIDERAÇÕES ACERCA DAS RACIONALIDADES TÉCNICA, PRÁTICA E DA REFLEXÃO NA FORMAÇÃO DE PROFESSORES. Cad. da Pedagog. **15**(33) (2021)
29. Cunha, A.M.: MODELOS DE FORMAÇÃO DOCENTE: movimentos e reflexões para umaabordagem teórico-prática. Formação@ Docente **13**(1), 150–170 (2021)
30. Kennedy, E.: Blended learning in teacher education & training: findings from research & practice. Brussels, Belgium Eur. Sch. Retrieved May, vol. 18, p. 2021 (2021)
31. Martín-Núñez, J.L., Bravo-Ramos, J.L., Sastre-Merino, S., Pablo-Lerchundi, I., Caravantes Redondo, A., Núñez-del-Río, C.: Teaching in secondary education teacher training with a hybrid model: students' perceptions. Sustainability **14**(6), 3272 (2022)
32. Burns, M.: Distance Education for Teacher Training: Modes, Models, and Methods. Educ. Dev. Center, Inc. (2023)
33. Grimes, G., Boyle, F., Noctor, M.: Online learning standards: steps to introduce a distributed leadership approach to training teachers for online teaching and learning. All Irel. J. High. Educ. **15**(3) (2023)
34. Sparks, D., Loucks-Horsley, S.: Models of staff development. Handb. Res. Teach. Educ. **3**, 234–250 (1990)
35. Calderhead, J., Shorrock, S.B.: Understanding teacher education: case studies in the professional development of beginning teachers. Psychology Press (1997)
36. Clarke, D., Hollingsworth, H.: Elaborating a model of teacher professional growth. Teach. Teach. Educ. **18**(8), 947–967 (2002)
37. Villegas-Reimers, E.: Teacher professional development: an international review of the literature. International Institute for Educational Planning Paris (2003)
38. Gaible, E., Burns, M.: Using Technology to Train Teachers: Appropriate Uses of ICT for Teacher Professional Development in Developing Countries. Online Submission (2005)
39. Lumpe, A.T.: Research-based professional development: teachers engaged in professional learning communities. J. Sci. Teach. Educ. **18**(1), 125–128 (2007)
40. Sachs, J.: Aprender para melhorarou melhorar a aprendizagem: O dilema do desenvolvimento profissional contínuo dos professors. Aprendiz. e Desenvolv. Prof. Profr. Context. e Perspetivas, pp. 99–118 (2009)
41. Gonçalves, T., Gomes, E.: Re-imaginar o desenvolvimentoprofissional contínuo de professores: O projecto 10X10 da Fundação Calouste Gulbenkian. Medi@ ções **2**(2), 63–80 (2014)
42. Daly, C., Pachler, N., Pelletier, C.: Continuing professional development in ICT for teachers: a literature review (2009)
43. Matzat, U.: Reducing problems of sociability in online communities: integrating online communication with offline interaction. Am. Behav. Sci. **53**(8), 1170–1193 (2010)
44. Matzat, U.: Do blended virtual learning communities enhance teachers' professional development more than purely virtual ones? A large scale empirical comparison. Comput. Educ. **60**(1), 40–51 (2013)

45. McConnell, T.J., Parker, J.M., Eberhardt, J., Koehler, M.J., Lundeberg, M.A.: Virtual professional learning communities: teachers' perceptions of virtual versus face-to-face professional development. J. Sci. Educ. Technol. **22**, 267–277 (2013)
46. Vangrieken, K., Meredith, C., Packer, T., Kyndt, E.: Teacher communities as a context for professional development: a systematic review. Teach. Teach. Educ. **61**, 47–59 (2017)
47. Lantz-Andersson, A., Lundin, M., Selwyn, N.: Twenty years of online teacher communities: a systematic review of formally-organized and informally-developed professional learning groups. Teach. Teach. Educ. **75**, 302–315 (2018)
48. Kimmelmann, N., Lang, J.: Linkage within teacher education: cooperative learning of teachers and student teachers. Eur. J. Teach. Educ. **42**(1), 52–64 (2019)
49. Kitchenham, A., Chasteauneuf, C.: Teacher supply and demand: issues in northern Canada. Can. J. Educ. **33**(4), 869–896 (2010)
50. Coutinho, C.P., Lisbôa, E.S.: Perspetivandomodelos de formação de professores que integram as TIC nas práticas letivas: um contributo para o estado da arte (2011)
51. Philipsen, B., Tondeur, J., McKenney, S., Zhu, C.: Supporting teacher reflection during online professional development: a logic modelling approach. Technol. Pedagog. Educ. **28**(2), 237–253 (2019)

Gender Bias in Tech – Young People's Perception of STEM in Portugal

Helena Elias[1] ⬚, Isabel Pedrosa[1,2(✉)] ⬚, and Maristela Holanda[3] ⬚

[1] Polytechnic Institute of Coimbra, Coimbra Business School, Quinta Agrícola - Bencanta, 3045-231 Coimbra, Portugal
ipedrosa@iscac.pt
[2] Polytechnic Institute of Porto, CEOS.PP, Porto, Portugal
[3] Texas A&M University, College Station, TX, USA

Abstract. Technological advancement is molding the future decisions of young individuals, sparking an increased interest in STEM (Science, Technology, Engineering, and Mathematics) disciplines. Nevertheless, concerns about gender inequality persist, manifesting themselves through stereotypes and gender perceptions. Ensuring equal opportunities is of utmost importance. In this paper, the following inquiries were addressed: "What is the influence of the technological evolution on the preferences and outlook of young individuals regarding STEM fields?" and "How is gender inequality manifested within this context, leading to differing responses among male and female audiences in Portugal?".

The findings underscore the significance of actively promoting gender diversity across all STEM areas. Encouraging more women to consider STEM careers and establishing awareness programs to challenge gender stereotypes are crucial steps. Despite the progress that has been made, the perception that certain professions are still associated with a specific gender continues to persist, emphasizing the ongoing necessity for education and awareness concerning gender equality in STEM fields.

This study contributes to our understanding of the impact of emerging technologies on the choices made by young individuals in STEM fields and on the enduring issue of gender inequality. It emphasizes the imperative to foster gender equality and cultivate inclusive environments within the STEM domain.

Keywords: Gender Inequality · Stereotype · STEM (Science Technology Engineering and Mathematics) · Women · Young people · Perception

1 Methodology - Systematic Literature Review

The principal objective of this review was to compile and present information of relevance for contextualizing the theme of gender inequality in STEM fields. The collection and analysis of articles aimed to procure precise and verifiable data, subsequently subjected to comparison with the outcomes obtained from an online questionnaire administered to students. This undertaking aimed to contribute to a deeper understanding of students' perceptions regarding gender disparities in STEM fields. To address the central

© The Author(s), under exclusive license to Springer Nature Switzerland AG 2024
Á. Rocha et al. (Eds.): WorldCIST 2024, LNNS 988, pp. 120–130, 2024.
https://doi.org/10.1007/978-3-031-60224-5_13

inquiry pertaining to gender inequality in STEM, particularly focusing on Technology, a Literature Review was conducted employing the Systematic Literature Review (SLR) methodology following the PRISMA (Preferred Reporting Items for Systematic Reviews and Meta-Analyses) guidelines.

2 Systematic Literature Review

Gender inequality in the field of Technology is an issue of extreme relevance today, as women have historically been marginalized in their participation in the fields of Technology and Science. In this regard, the present study aims to analyze the perceptions of young individuals who are nearing the end of their educational journey regarding STEM education (Science, Technology, Engineering, and Mathematics) in Portugal.

The digital revolution has wielded a profound influence on our lives, ushering in opportunities while concurrently presenting substantial challenges. Ensuring equal opportunities in the labor market, fostering equitable treatment in the workplace, and striving for gender balance within the digital realm are pivotal. This pursuit not only augments the European Union's economic growth in terms of GDP but also embodies a matter of justice for all exceptionally talented women choosing careers in STEM fields, as noted in [1].

As articulated in [1] and [2] women have made considerable progress in realms like higher education, politics, and the workforce since the early 20th century. Nonetheless, they continue to confront substantial barriers within STEM domains, traditionally dominated by men.

As highlighted in [3] gender inequality in the sciences remains a prevalent topic in the 21st century. Despite a marked expansion in studies addressing this matter in recent decades, the endeavor persists to convince society of the criticality of diversity and its correlation with enhanced scientific output. Overcoming this challenge may well hinge on the imperative need for data substantiating efficacious proposals to reshape the academic milieu. Despite strides towards gender parity, women remain significantly underrepresented in leadership roles within STEM, and progress in this regard has been sluggish since the 1960s. This phenomenon, often termed the "leaky pipeline", underscores the departure of numerous women from STEM careers due to specific gender-related hurdles, such as biased research assessment metrics.

The goal is to understand how these young individuals are responding to campaigns promoting female involvement in these fields. The persistence of gender inequality in STEM fields is widely recognized and continues to be a concern for educators, teachers and the scientific community, as mentioned in [4]. This disparity is not just restricted to female participation, but also to perceptions and attitudes towards these disciplines. Even though the gender gap in STEM achievement is closing, there are still regions of the world where disparities persist.

The underrepresentation of women in STEM fields has significant implications for educational and socioeconomic gender disparities. Deeply understanding the reasons behind this phenomenon is one of motivations for this study.

The simplistic view that attributes the gender disparity in STEM to differences in mathematical skills has lost popularity, but still persists in the literature. This classic approach makes it difficult to understand the complex reasons for female underrepresentation in STEM, as mentioned by [5]. Current research and Portuguese-led initiatives in the STEM field highlight the importance of preparing young people for global competitiveness, emphasizing the need for integrated STEM education for multidisciplinary understanding and comprehensive skills.

3   Data Collection and Processing

As a general objective, intend to analyze the topic through data collection involving surveys of 9th and 12th classes of Professional and Scientific-Humanistic Education, from the Ponte de Sor Group, with the theme "Gender inequality in Technology - young people's perception of STEM in Portugal". Therefore, the main objective will be to identify gender differences in relation to the preferred training routes and the future expectations of these young people in relation to the STEM areas, placing greater emphasis on the technological area. In this context, it is intended that the visualizations of the insights resulting from this study can contribute to facilitating the understanding and sharing of the conclusions with the community and to understand in a graphic and interactive way the research and answer the key questions: "Does the evolution of young people in new technologies have an impact on their future choices and opinions regarding STEM areas? How is gender inequality manifested in this context? (depending on answers given by female or male audience)"First and foremost, it is important to highlight that data analysis is a process that examines data collected from specific questions and applies analytical techniques to extract meaningful information. The data analysis process involves various stages, such as data preparation, exploratory data analysis, statistical analysis, relationship analysis, and trend analysis.

As mentioned in [6] factors related to location, school ownership, and gender are considered important indicators when assessing educational effectiveness in terms of quality and equity. The following data analysis is based on the results of a sample, which is representative of the entire population of the Ponte de Sor municipality in Portalegre, Portugal. In this context, the data from the questionnaires for the secondary education population of the Agrupamento de Escolas de Ponte de Sor, which does not exhibit sampling errors, will be analyzed.

4   Survey Design

The use of questionnaires as a data collection instrument allowed for a more comprehensive insight into the opinions, perspectives, and experiences of the study group. Additionally, this technique enabled the quantification and systematization of responses, facilitating the interpretation and statistical analysis of results.

Therefore, it can be stated that the use of questionnaires played a crucial role in this research, providing a deeper understanding of the topic at hand and significantly contributing to the success and rigor of this study.

Developing an effective questionnaire, in this case on the LimeSurvey platform, involved several crucial steps to ensure that the questions were clear, relevant and capable of eliciting the desired information from respondents.

As mentioned in [2] and in [3] regardless of the chosen tool for creating the questionnaire, within each stage there are steps that should be followed to achieve the desired outcome, including:

1. **Development:** To create an effective questionnaire, establish research objectives, choose question types, organize questions, select response formats, and ensure clarity and impartiality.
2. **Pre-Testing:** Prior to launch, choose a survey platform, configure the questionnaire, customize its appearance, conduct a pre-test with a small sample to address comprehension issues, and maintain a conducive environment.
3. **Administration:** Publish and distribute the questionnaire through email, social media, or a website, and monitor responses for adjustments to boost the response rate.
4. **Closing and Analysis of Responses:** Close the questionnaire after achieving the desired sample or time limit, analyze data with statistical tools, and draw conclusions based on the collected information.

5   Survey Data

The development of the survey questions was based on the following papers: [7–11]. This contribution was of utmost importance since a significant portion of the reviewed papers consulted did not contain questionnaires that could serve as a reference for creating the questions to the survey for this study.

The survey was administered to students across various educational levels at the conclusion of the academic year. Specifically, it targeted students in the 9th year, 12th year of Scientific-Humanistic Education, and 12th year of Professional Education classes within the Municipality of Ponte de Sor, located in Central Portugal. A total of 61 students participated in the survey, encompassing 29.51% girls, 42.63% boys, and 27.87% who did not disclose their gender. The survey comprised a series of direct inquiries pertaining to the subject matter and was conducted online.

6   Data Preparation with Power Query and Power BI

Power Query is a data transformation and preparation tool developed by Microsoft. It is an ETL (Extraction, Transformation, Loading) tool used to extract data from sources, process it, and load it into one or more target systems, as referenced in [4].

Main steps taken in Power Query for processing data extracted in Excel from LimeSurvey:

- Step 1 - Promote header rows: The "Promote Header" feature was used to transform the first row of the table into column headers.
- Step 2 - Replace values for blank and N/A rows: The "Replace Values" option was used to find blank or N/A values and replace them with desired values, such as "N/A" (Not Available) or other relevant terms.

- Step 3 - Create an ID for all respondents to identify each response and to anonymize each response: A custom column was added with a row counter to create unique IDs for each respondent using the "Add Custom Column" function.
- Step 4 - Use filters: Filters were used to exclude responses that were not relevant for analysis. For example, respondent number 12 was the only one who provided a different (course) response at the 9th-grade level, making it irrelevant for the analysis.
- Step 5 - Create new tables: Quick measure tables were created to present questions with multiple subparts and several possible responses for each subpart.
- Step 6 - Table relationships: star schema model, as all tables are related to the "base data" table: after completing all the previous steps, relationships were configured between the tables according to the star schema model. This involved creating relationships between the "base data" table and the new tables created in steps 5 and 6, using columns that serve as relationship keys.

After performing all these data transformations procedures, Power Query allowed the creation of a star schema model in Power BI. This ensures that tables are related properly, enabling effective and efficient analysis.

Microsoft Power BI proved to be a valuable tool for the current data analysis, providing advanced visualization and analysis capabilities. With its intuitive interface and report generation capabilities, we were able to explore survey results in a more detailed and effective manner. This enabled us to identify trends, patterns, and relevant insights that significantly contributed to the understanding of the research topic.

For each questionnaire-related question, specific queries were formulated for which each dashboard should provide answers. This allowed for the effective and targeted visualization and analysis of the collected data. The use of dashboards assisted in extracting insights and important information from the questionnaire responses.

7 Data Analysis

Table 1 displays the main questions created resulting from the responses to the questionnaires as well as the presentation of the respective dashboards that were created to obtain these responses. An analysis is also carried out on each dashboard.

It is important to emphasize that this approach aims to distinguish between the responses given by female and male students.

Table 1. Dashboard analysis

Survey Question: What Scientific-Humanistic major do you attend or intend to attend?
Question the dashboard should answer - What percentage of boys attend/intend to attend the Science and Technology area compared to the percentage of girls? Is the result the same for Languages and Humanities?

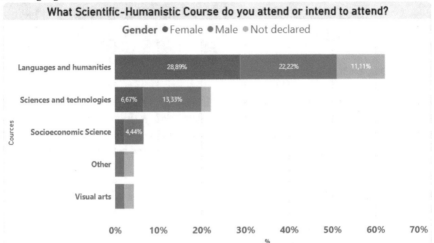

Figure 1 - Preference for study areas

Analysis: An evident disparity exists in the selection of preferred fields rather than a pronounced gender-based difference. The *Figure 1* indicates that 28.89% of girls favor Humanities and Languages, while only 6.67% show preference for the Science and Technology domain. In contrast, boys display a relatively narrower gap between the two areas, with 22.22% favoring Languages and Humanities compared to 13.33% opting for Science and Technology.

Survey Question: Would you consider taking a higher education course related to IT?
Question the dashboard should answer - What percentage of boys and girls want to take a higher education course related to IT? Depending on this answer, what reasons led the girls and boys to make that decision? Are there boys and girls who mentioned the same reasons?

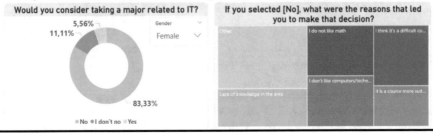

(*continued*)

Table 1. (*continued*)

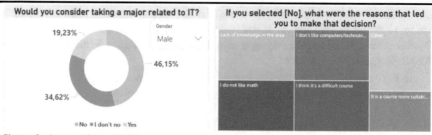

Figure 2 - Interest in major IT

Analysis: A mere 5.56% of girls express consideration for pursuing a major in IT, with their reasons predominantly extending beyond the available choices. Instead, their rationale leans towards a general disinterest in the field. Conversely, boys exhibit a higher percentage of interest, amounting to 19.23%. Notably, among the disinterested boys, the reasons provided align with the options offered, including an aversion to studying.

The underrepresentation of girls in IT-related majors is a prevalent phenomenon, attributable to various factors such as gender stereotypes, the absence of female role models, misconceptions about the industry, and a lack of encouragement.

When girls cite a lack of interest in IT fields as a deterrent to pursuing related courses, this tendency might be linked to their limited exposure to opportunities in the field and the scarcity of female role models. Additionally, pervasive gender stereotypes might foster a self-perception among girls that the IT domain isn't suitable for them, further influencing their career choices.

Survey Question: How often do you use a computer?

Question the dashboard should answer - Do girls use PCs more or less frequently than boys? (Draw conclusions that the use of the PC influences/does not influence the choice of paths in the Technology area.)

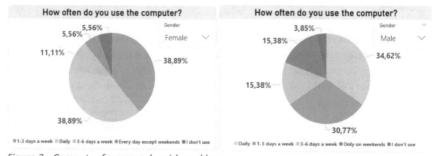

Figure 3 - Computer frequency by girls and boys

Analysis: In this context, there exists a parallel in the frequency of computer usage between girls and boys. Consequently, it is not conclusive to infer that computer usage directly influences interest in this field. *Figure 3* illustrates that 38.89% of girls use computers daily compared to 34.62% of boys, indicating a relatively minor difference. Moreover, the percentage of girls not utilizing computers stands at 5.56%, slightly higher than the 3.85% of boys who do not engage with computers.

(*continued*)

Table 1. (*continued*)

Survey Question: For each of the STEM professions, indicate your gender perception.
Question the dashboard should answer - For each of the areas, is there a lot of divergence in the responses of boys compared to girls regarding whether they are male or female STEM professions?

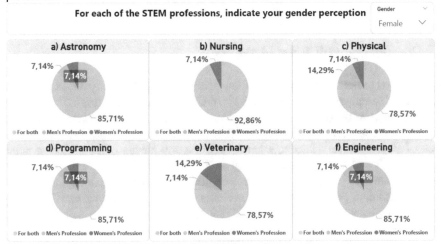

Figure 4 - Gender in STEM professions - female perspective

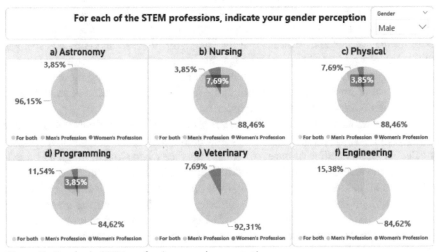

Figure 5 - Gender in STEM professions - male perspective

Analysis: It is evident from *Figure 4* and *Figure 5* that in the majority of professions, both boys and girls indicated that these occupations were suitable for both genders. In terms of gender stereotyping, Engineering was perceived as masculine by 15.38% of boys, while Astronomy was perceived similarly by 3.85% of boys. In contrast, Veterinary practice was stereotyped as masculine by 7.69% of boys, and Nursing was considered feminine by 7.14% of girls.

8   Discussion of Results

The discrepancy between the questionnaire results and the bibliographic data can be explained by differences in data collection periods and research focus. While the questionnaire focused on 9th and 12th-grade students who were already making university decisions, the bibliographic source [7] one of the main ones used, included information from younger students who were still a long way from making that decision.

9   Comparing with Related Works

It is essential to note that the findings drawn from this data may vary depending on the context and sample specifics. As such, it is crucial to interpret these results with care and take into account the differences in data collection methods and the samples involved. In summary, both sets of data underscore the need to promote an equitable and inclusive view of all professions related to science, encouraging students of all genders to pursue their interests and talents regardless of gender stereotypes. The data collected through the questionnaires identified career choices, familiarity with technology, the level of logical reasoning, primary reasons for avoiding careers in the Science and Technology field, self-efficacy related to computer literacy, and issues related to gender bias and barriers [9].

These conclusions align with what the article [5] examined, particularly the persistence of the "myth of the male mathematician" and the associated "myth that females are not good at mathematics." This study highlights the complex interplay of social and cultural forces that sustain these stereotypes. By combining recent findings from various research areas, it becomes evident that the longevity of these myths results from the additive influence of two independent cognitive biases related to gender and mathematical stereotypes. This study emphasizes the importance of addressing these stereotypes holistically to promote a more accurate and equitable understanding of mathematical skills in all individuals, regardless of gender.

Furthermore, as mentioned in [12] understanding girls' perceptions of self-efficacy in the field of computing and technology and its impact on career choices is essential for a deeper understanding of the experiences they face when striving for greater gender diversity in the technology field. Gender differences play a significant role in the career decisions of adolescent girls, influencing their choices of higher education courses and careers in computing and technology. Addressing this issue from the perspective of Career Development Psychology and considering the role of self-efficacy can help identify strategies to develop more inclusive support systems free from gender stereotypes in non-traditional fields like technology.

As highlighted in [8], despite women being pioneers in computing, the interest of high school female students in computing/technology and their participation in the academic and professional computing environment have been decreasing over time. This aligns with the conclusions drawn from the statistics, emphasizing the importance of providing clear information about careers in technology to young people and clarifying existing perceptions and ambiguities.

10   Conclusion

This study can serve as a guide for future initiatives and policies aimed at reducing gender disparities in this field and encouraging more young people to explore this subject. In conclusion, the aim is to address the key questions, namely the Research Questions: "Does the evolution of young people in new technologies have an impact on their future choices and opinions regarding STEM areas? How is gender inequality manifested in this context? (depending on answers given by female or male audience)".

Based on the conclusions drawn from the analysis of young people's responses regarding STEM fields and gender perceptions, we can infer the following in response to the research questions: 1) Impact of New Technologies on Future Choices - there appears to be a correlation between young people's access to and familiarity with new technologies and their interest in STEM fields. Those who are more involved in technology tend to express greater interest and confidence in these areas. This suggests that new technologies play a significant role in shaping young people's educational and career choices; 2) Gender Perceptions and Inequality in STEM Field - significant differences in gender perceptions related to STEM fields were observed. On average, girls tend to express more negative perceptions and less confidence in mathematics and science than boys. Additionally, the research identified persistent gender stereotypes, with some professions still being viewed as more suitable for a specific gender (e.g., nursing as a female profession and engineering as a male profession); 3) Gender Inequality and Stereotypes in STEM - differences in gender perceptions can contribute to gender inequality in STEM fields. Girls may be discouraged from pursuing careers in these fields due to negative perceptions or lack of confidence. At the same time, boys may be less encouraged to explore traditionally female fields, such as nursing.

While the results indicate progress towards a more inclusive outlook, there is still work to be done. The fact that some respondents still perceive certain professions as more masculine than feminine underscores the ongoing need for education and awareness regarding gender equality in STEM. As future work, it may also be interesting to apply questionnaires at the level of various teaching cycles, in different regions of Portugal (rural, urban…), which allow analyzing students' sense of belonging within the academic community and studying to what extent the subjects influence student performance and self-efficacy as well as identifying areas where it is necessary to implement measures to improve gender equality at school, to help address problems related to sexism and discrimination.

The statistics obtained through the questionnaires will contribute for decision-makers such as educators, employers, and policymakers who wish to promote gender equality in STEM. They can use this data to guide recruitment initiatives, educational programs, and policies aimed at creating a more inclusive environment across all STEM fields.

Acknowledgement. This work received financial support from the Polytechnic Institute of Coimbra within the scope of Regulamento de Apoio à Publicação Científica dos Estudantes do Instituto Politécnico de Coimbra (Despacho n.º 5545/2020).

References

1. Rodríguez-lozano, P., et al.: Women in limnology: from a historical perspective to a present-day evaluation, November 2021, pp. 1–14 (2022). https://doi.org/10.1002/wat2.1616
2. Charlesworth, X.T.E.S., Banaji, M.R.: Gender in Science, Technology, Engineering, and Mathematics: Issues, Causes, Solutions, vol. 39, no. 37, pp. 7228–7243 (2019)
3. Shirai, L.T., et al.: Brazilian female researchers do not publish less despite an academic structure that deepens sex gap, pp. 1–17 (2022). https://doi.org/10.1371/journal.pone.0273291
4. Reilly, D., Neumann, D.L., Andrews, G.: Investigating gender differences in mathematics and science: results from the 2011 trends in mathematics and science survey, pp. 25–50 (2017). https://doi.org/10.1007/s11165-017-9630-6
5. Girelli, L.: What does gender has to do with math? Complex questions require complex answers. J. Neurosci. Res. (2022). https://doi.org/10.1002/jnr.25056
6. Jakaitienė, A., Želvys, R., Vaitekaitis, J., Raižienė, S., Dukynaitė, R.: Centralised mathematics assessments of Lithuanian secondary school students: population analysis. Inform. Educ. 20(3), 439–462 (2021). https://doi.org/10.15388/infedu.2021.18
7. Holanda, M., Ramos, G., Mourão, R., Araujo, A., Emília, M., Walter, T.: Percepção das Meninas do Ensino Médio sobre o Curso de Computação no Distrito Federal do Brasil
8. Ferreira, T., Dias, E.: A influência de uma ação de inclusão no interesse das alunas de ensino médio em cursar Computação na Universidade Federal de Goiás, pp. 164–168 (2020). https://doi.org/10.5753/wit.2019.6730
9. Moura, A.F.S.A., Tavares, T.H.C., Mattos, G.d.O., Moreira, J.A.: Incentivando alunas do Ensino Médio a Ingressarem em Carreiras de Ciência e Tecnologia na Paraíba, pp. 1–5 (2020). https://doi.org/10.5753/wit.2018.3380
10. Zuazu, I.: Graduates' opium? Cultural values, religiosity and gender segregation by field of study. Soc. Sci. 9(8) (2020). https://doi.org/10.3390/SOCSCI9080135
11. Baselga, S.V.: Drama-Based Activities for STEM Education: Encouraging Scientific Aspirations and Debunking Stereotypes in Secondary School Students in Spain and the UK, pp. 173–190 (2022)
12. Figueiredo, K.D.S., Maciel, C.: A autoeficácia no desenvolvimento de carreira e sua influência na diversidade de gênero na computação. Rev. Educ. Pública 27(65/1), 365 (2018). https://doi.org/10.29286/rep.v27i65/1.6586

Digital Technology in Higher Education: Exponential Function Case

Edwin Cristian Julián Trujillo[1]([📧]) [iD], Jorge Luis Vivas-Pachas[2] [iD], and Jesús Victoria Flores Salazar[3] [iD]

[1] Engineering School, Universidad San Ignacio de Loyola, La Molina, 15024 Lima, Peru
edwin.julian@usil.pe
[2] Universidad de Lima, Lima, Peru
jvivas@ulima.edu.pe
[3] Pontifical Catholic University of Peru, San Miguel, 15088 Lima, Peru
jvflores@pucp.pe

Abstract. This study is part of a research project seeking to characterize the mathematical work of higher education students as they engage with the concept of exponential functions within a didactic sequence facilitated by digital technology. Herein, we discuss one of the tasks proposed in the sequence and assess the expected results within the framework of the Mathematical Working Spaces (MWS) theory.

Keywords: Exponential Function · Mathematical Working Space · GeoGebra

1 Introduction

This research study, a comprehensive part of an ongoing project, is qualitative in nature and incorporates an experimental segment concentrating on higher education students. Our interest lies in emphasizing the potential of digital technology, specifically the GeoGebra software, in solving tasks related to exponential functions to characterize student mathematical work. The referenced research is in line with our investigative interests.

We conducted a literature review to gain a deeper understanding of students' mathematical work when tackling tasks related to exponential functions. Brucki (2011) assessed how an activity linked to exponential functions aids learning. The activity involved Brazilian high school freshmen and aimed to connect the algebraic model of the exponential function to the general term of the geometric progression. The author concluded that extra mathematical activities enhance learning when they incorporate ideas to support the concept of the exponential function and/or geometric progression.

In addition, Sureda and Otero (2013) delved into the conceptualization of learning exponential functions, emphasizing the necessity for instructional design integrating multiple representation systems. To this end, they meticulously planned, designed, implemented, and analyzed situations encompassing five different representation systems to reconstruct the conceptual field of the exponential function for fourth-year high school students. However, the authors stated that their research did not offer the necessary support to claim that the conceptualization process invariably goes through each of the stages under consideration. Nonetheless, they concluded that the clarification, discussion, and formalization of concepts in each representation system are crucial for the transition from the linear to exponential stage.

Vivas (2020) analyzed the mathematical work of higher education students when solving a task on exponential functions. The experimental phase of this research was unfolded in a class session with first-cycle humanities students at a private university in Lima, Peru. The task comprised two questions grounded in problems related to exponential functions, demanding the use of algebraic and graphical representations. In this study, the elements of the Mathematical Working Space theory and the personal mathematical work analysis method were used. The analysis disclosed the activation of semiotic, instrumental, and discursive genesis, alongside semiotic-instrumental, semiotic-discursive, and instrumental-discursive vertical planes. This identified that the mathematical work of the student falls within the Analytic-Geometric Analysis paradigm.

Hence, this study seeks to assess the potential actions of higher education students to characterize their mathematical work in tasks related to exponential functions, facilitated by digital technology.

2   The Mathematical Working Space (MWS)

The study includes elements from the MWS theory, as elucidated by Kuzniak and Richard (2014), outlining that MWS comprises epistemological and cognitive planes. The epistemological plane encompasses three key components: representamen, artifact, and theoretical referent. Conversely, within the cognitive plane, three distinct processes are identified: visualization, construction, and testing. The interconnection of these planes is facilitated through the activation of three genesis: semiotic, instrumental, and discursive. This activation, in a cascade, enables the activation of vertical planes: semiotic-instrumental

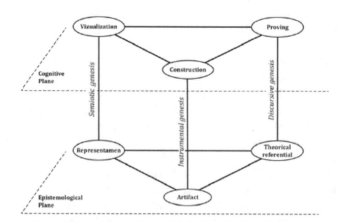

Fig. 1. MWS model. **Source:** Kuzniak, Tanguay and Elia [5]

[Sem-Ins], semiotic-discursive [Sem-Dis], and instrumental-discursive [Ins-Dis] (see Fig. 1).

Our research is qualitative, emphasizing the analysis of the specific didactic phenomenon under consideration. To assess student mathematical work, we relied on the method developed by Kuzniak and Nechache (2018). Therein, they assert that mathematical action is objectified from written or oral discourse, and an episode comprises a series of mathematical actions performed by the student to fulfill a task. This method comprises two stages. The initial stage involves identifying mathematical actions within production, grouping them into episodes, and interpreting them through the lens of the MWS, which includes identifying the activation of genesis and vertical planes. The subsequent stage aims to offer a comprehensive overview of the circulation of mathematical work, based on the MWS framework. This paper focuses on the assessment derived from the first stage of the method.

Regarding artifacts and acknowledging the research's reliance on digital technology, it is essential to specify what qualifies as a digital artifact for the teaching and learning of mathematics, according to MWS. Thus, Salazar, Gaona, and Richard (2022) define a digital artifact as a set of propositions executable by an electronic machine with historical intelligence and relative epistemological validity. Furthermore, the discursive genesis leverages the properties of the theoretical reference system to support mathematical reasoning.

Concerning tasks, Gaona (2022) underscores their significance, noting that even when they are not part of MWS, they play a crucial role in driving the activation of mathematical work.

3 The Task

The following task is integrated into the didactic sequence, comprising a file from the GeoGebra software and a response sheet containing the provided instructions and questions (see Fig. 2).

Open the GeoGebra file, use the C and "a" sliders to obtain the graphical representation of different exponential functions for $f(x) = C.a^x$, $x \in \mathbb{R}$.

Answer the following questions, justifying your results:

Question 1

Is there any element of the graphical representation of the function that allows us to determine the value of C? If so, mention how this element related to the value of C.

Question 2

Assess the following cases and answer:

Case 1: Choose any value for a, between 0 and 1 $(0 < a < 1)$ and use the C slider. For which values, positive or negative, is the function increasing and decreasing?

Case 2: Choose any value for a, greater than 1 $(a > 1)$ and use the C slider. For which values, positive or negative, is the function increasing and decreasing?

The objective of this task is for students to identify and justify the y-intercept and monotony of an exponential function. To facilitate this, a GeoGebra applet is supplied, enabling the graphical depiction of exponential functions using two sliders.

In Fig. 2, the graphical representation of the exponential function in the applet, indicating that the y-intercept can have either a positive ordinate (a) or a negative ordinate (b).

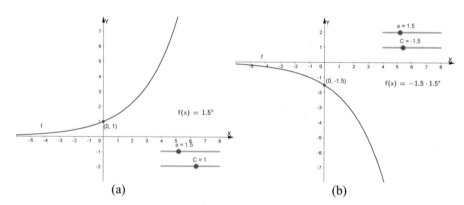

Fig. 2. Intercept with the Y-axis (a and b). **Source:** Prepared by the authors

In Fig. 3 below, the graphical representation of the increasing exponential function is displayed.

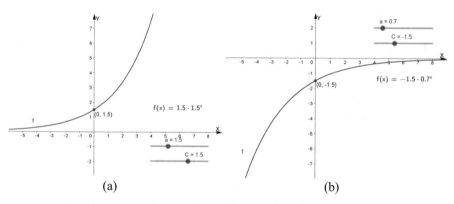

Fig. 3. Increasing function (a and b). **Source:** Prepared by the authors

In Fig. 4 below, the graphical representation of the decreasing exponential function is displayed.

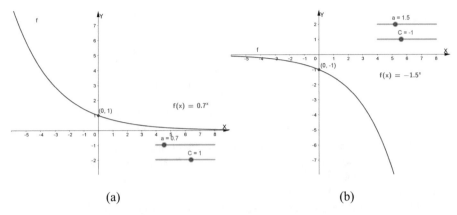

(a)   (b)

Fig. 4. Decreasing function (a and b). **Source:** Prepared by the authors

4 Analysis of Expected Mathematical Production

The analysis of the expected mathematical output for the task, based on the MWS theory and according to the initial stage of the Kuzniak and Nechache method (2018), is discussed below.

To address Question 1, manipulation of the "C" slider is expected to identify the value of C as the ordinate of the intercept with the Y-axis. Within the framework of the ETM, the "C" slider serves as a digital artifact, engaging in a construction process that facilitates the acquisition of the graphical representation of f. Consequently, the instrumental genesis is set into motion. Then, the intercept with the Y-axis functions as a representamen, undergoing a visualization process to pinpoint the value of the ordinate of the intercept with the value of C. This highlights the activation of the semiotic genesis and, consequently, the activation of the vertical plane [Sem-Ins].

In Fig. 5, the use of components from the epistemological plane (representamen and digital artifact) are displayed in the applet, for Question 1.

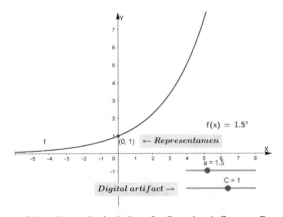

Fig. 5. Components of the epistemological plane for Question 1. **Source:** Prepared by the authors

In relation to Question 2, the analysis for Case 1 ($0 < a < 1$) is presented. To respond to this case, it is expected that manipulating the "a" slider will allow identifying how the monotony of f changes with the sign of C. In terms of the MWS, the "a" slider is used as a digital artifact to drive the activation of the discursive genesis since the condition from the statement $0 < a < 1$ is used as a theoretical referent. A test is conducted to justify the choice of "a." Subsequently, the "C" slider is employed as a digital artifact to undergo a construction process that enables obtaining the graphical representation of f. This activates the instrumental genesis and, consequently, the vertical plane [Ins-Dis]. Finally, the definition of an increasing (or decreasing) function, expressed in the graphical representation of f, serves as a theoretical referent to conduct a pragmatic test justifying whether the value of C should be positive (or negative), leading to the activation of the discursive genesis.

In Fig. 6, the use of components from the epistemological plane (digital artifacts) present in the applet is shown for Question 2 - Case 1.

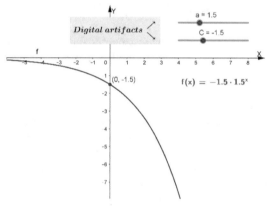

Fig. 6. Epistemological Plane Components for Question 2-Case 1. **Source:** Prepared by the authors

5   Conclusions

We believe that the task contributes to characterizing the student's mathematical work by facilitating the activation of the semiotic, instrumental, and discursive genesis, along with the vertical planes [Sem-Ins] and [Ins-Dis]. This becomes evident when adjustments are made to the graphical representation of the exponential function using the sliders provided in the GeoGebra applet, enabling the observation of changes in its intercept with the y-axis and monotony.

Furthermore, obtaining the graphical representation of various exponential functions is achievable solely through the manipulation of the GeoGebra sliders. This mathematical action highlights the activation of the instrumental genesis. Subsequently, the graphical representation of the exponential function in the applet facilitates the identification of its intercept with the y-axis, triggering the activation of the semiotic genesis. Similarly, addressing questions regarding the monotony of the exponential function in the applet can be translated as the activation of the discursive genesis. In this way, these actions represent processes that activate the vertical planes [Sem-Ins] and [Ins-Dis].

Consequently, we assert that a didactic sequence involving tasks mediated by digital technology aids in characterizing the student's mathematical work since it allows the analysis of their potential mathematical actions, interpreted through the activation of the semiotic, instrumental, and discursive genesis, as well as the vertical planes [Sem-Ins] and [Ins-Dis].

Acknowledgement. We appreciate the collaboration from the Universidad San Ignacio de Loyola (USIL) for the English translation of this research.

References

1. Brucki, C.: O uso de Modelagem no ensino de função exponencial (2011). https://tede2.pucsp. br/handle/handle/10900
2. Gaona, J.: Diseño de tareas en un sistema de evaluación en línea, una mirada desde la teoría de Espacios de Trabajo Matemático. [Task design in an online evaluation system, a look from the theory of Mathematical Working Spaces]. PädiUAQ: Revista de Proyectos y Textos Académicos en Didáctica de las Ciencias y la Ingeniería **5**(10), 2–18 (2022). https://revistas. uaq.mx/index.php/padi/issue/view/67
3. Kuzniak, A., Richard, P.: Espacios de trabajo matemático. [Mathematical Working Spaces] Puntos de vista y perspectivas. Relime **17**(4), 9–10 (2014). https://doi.org/10.12802/relime.13. 1741a
4. Kuzniak, A., Nechache, A.: Una metodología para analizar el trabajo personal de los estudiantes en la teoría de los espacios de trabajo matemático. [A methodology to analyze the personal work of students in the theory of mathematical working spaces] In: Montoya (Presidence), E. (ed.) Espacio de Trabajo Matemático. Lecture part of the Sixth International MWS Conference, Valparaíso, Chile (2018)
5. Kuzniak, A., Tanguay, D., Elia, I.: Mathematical working spaces in schooling: an introduction. ZDM-Math. Educ. **48**(6), 721–737 (2016). https://doi.org/10.1007/s11858-016-0812-x
6. Flores Salazar, J.V., Gaona, J., Richard, P.R.: Mathematical work in the digital age. Variety of tools and the role of geneses. In: Kuzniak, A., Montoya-Delgadillo, E., Richard, P.R. (eds.) Mathematical Work in Educational Context. MEDE, vol. 18, pp. 165–209. Springer, Cham (2022). https://doi.org/10.1007/978-3-030-90850-8_8
7. Sureda, P., Otero, M.: Estudio sobre el proceso de conceptualización de la función exponencial. Educación Matemática [Study on the conceptualization process of the exponential function. Math. Educ.] **25**(2), 89–118 (2013). http://www.revista-educacion-matematica.com/pdf/doc umentos/REM/REM25-2/Vol25-2-4.pdf
8. Vivas-Pachas, J.: Trabajo matemático de estudiantes de humanidades en tareas sobre sobre función exponencial. [Mathematical work of humanities students in exponential function tasks] (2020). http://hdl.handle.net/20.500.12404/18104digi

Adapting Traditional Whiteboarding for Remote Education Using Real-Time Handwritten Content Detection System

Dawid Skurzok and Jerzy Nowiński[✉]

ShareTheBoard, From Poland With Dev, Czysta 10/5, 31-121 Kraków, Poland
jerzy@sharetheboard.com

Abstract. This paper introduces an innovative system designed for real-time detection and amplification of handwritten content on whiteboards, utilizing standard classroom equipment (webcam, laptop). Aimed at augmenting remote and hybrid education, the system employs client-side computing computer vision and machine learning algorithms. It recognizes handwritten content and identifies potential obstacles, such as individuals interacting with the board. This enables the system to overlay the amplified content onto the output video stream, ensuring that the teacher's body does not obstruct the view for remote learners. Consequently, this approach facilitates an engaging and interactive learning experience for students participating via video conferencing solutions. The paper discusses the technical aspects of the system and evaluates its effectiveness in enhancing the educational experience in remote learning scenarios.

Keywords: edge-computing · computer vision · remote education · handwritten content detection

1 Introduction

The COVID-19 pandemic accelerated the transition to remote and hybrid education in schools, universities, and private tutoring. This shift has highlighted significant challenges in the realm of knowledge transfer, particularly in the context of traditional teaching methodologies. Central to this issue is the reliance on handwriting and board-based teaching, a cornerstone of effective educational practices. Available applications, such as digital collaboration platforms, despite their advancements, fall short in emulating the ease, naturalness, and precision of traditional whiteboards removing a teacher's mimics and body language, which are crucial for engaging and effective teaching.

This paper introduces a novel system aimed at enhancing remote education. The main idea is to reintroduce in the digital environment the experience of classical whiteboarding providing teachers and remote students a familiar educational setting, complete with eye contact and natural body language. This is achieved through an application with a proprietary handwritten content detector that keeps whiteboard content clear and visible, overlaying over individuals interacting with the board in a video stream Additionally, the system features a separate application for remote participants, offering the ability to display and interact with digitized board content through digital annotations.

Emphasizing accessibility, the proposed solution does not require any additional or non-standard purchases, thereby enhancing its adaptability in educational settings.

2 Related Work

In addressing the challenges of remote work, various technological solutions have been developed. Hardware-based options, such as multimedia Smart Boards (e.g., [1]), are constrained by work area size, and suffer from the presenter's silhouette obstructing the view in remote conferences. Alternative hardware solutions like content cameras (e.g., [2]) offer content digitization but lack the capability for simultaneous teacher and board presentations. A notable commercial advancement is a content detection system [3] which enables overlaying detected content over the presenter.

3 System Architecture

3.1 Design Constraints

In addressing the challenges of handwritten content detection in real time in video streams, convolutional neural networks (CNNs) [4] were identified as the primary technique. Initial considerations involved server-side (backend) implementation; however, this approach was abandoned due to its intensive computational demands on CPU/GPU resources, leading to prohibitive infrastructure maintenance costs. To overcome these constraints, the focus shifted towards client-side computation, which mitigated the resource-intensive requirements of server-side processing.

3.2 Content Detector

Content and Obstacle Recognition Algorithms
The initial step in image processing involves limiting the full frame to a designated working area (e.g., the surface of a whiteboard) either marked by the user or automatically detected by a workspace detection algorithm. Such restriction accelerates computational processes as the speed of the content detection algorithm depends on the number of input pixels.

For content detection, the system employs one of two types of detectors. The first, based on classic computer vision algorithms, caters to laptops with lower hardware capabilities. The second utilizes convolutional neural networks (CNN) for recognition of the handwritten content. Figure 1, highlight the high-level structure of the CCN based algorithm.

Data Acquisition for Model Training and Verification
During the research, a variety of data sources were utilized to train the neural network model, including research team recordings, a small publicly available dataset "Access-Math: Whiteboard Video Summarization" [5] and data taken from publicly available

Fig. 1. High-Level Structure of the Content and Obstacle Recognition CNN Algorithm

educational video materials. The collected frames, totaling over 3500, encompassed various contents, including educational materials.

Verification and Accuracy of Handwritten Content Classification Algorithm
To verify the accuracy of the CCN algorithm developed for classifying handwritten content, a test dataset comprising 10% of all annotated images (303 frames) was randomly selected. The results for the model exceeded initial expectations as the content identification accuracy parameter reached 96.3%, a significant achievement considering the algorithm's satisfactory operational speed.

3.3 User Applications

The presenter application, designed for teachers conducting lectures with a physical whiteboard, processes the video feed from a webcam through a content detector. The content information extracted from this feed is vectorized [6], enhancing the visual quality of the results. The resulting SVG image is then distributed to all remote participants via a dedicated channel and concurrently overlaid on the video stream within the application's interface. Additionally, digital annotations added through the application by both the presenter and remote participants are layered on the video stream. The application's use case assumes that the resulting stream is shared with remote participants via a preferred video conferencing solution, utilizing the screen sharing functionality. This approach reduces infrastructure costs, concurrently allowing educators to maintain their established practices in utilizing favored communication tools with students.

Fig. 2. Image taken from the output stream of the application.

Figure 2 displays the presenter application's screen, highlighting clear, legible content layered over the teacher's image, along with digital annotations added by remote users.

Users connecting remotely or located in distant parts of a classroom can observe content in a clear, vector format using an independent application. Simultaneously, this solution enables remote interaction through a set of digital annotations.

4   Conclusions

The solution has been successfully tested in educational institutions across the USA, the Netherlands, and Canada. During the validation phase, the high precision of the detector was confirmed, demonstrating the application's effectiveness in facilitating remote and hybrid work with traditional mediums like whiteboards and flipcharts.

Thanks to various optimizations, the system operates efficiently on standard laptops, and webcams which are commonly used in educational settings. This compatibility greatly enhances the accessibility and applicability of the application, rendering it a practical and efficient tool suitable for diverse educational environments.

Acknowledgements. Research reported in this paper was supported by European Funds from the Smart Growth Operational Program 2014–2020 under the project: "Development and implementation of the innovative Share The Board app, using real-time digitization to easily save and share handwritten content with remote viewers".

References

1. Samsung Interactive Pro 75. https://www.samsung.com/us/displays/interactive/wm-series/samsung-interactive-pro-75-lh75wmbwlgcxza/. Accessed 20 Nov 2023
2. Logitech Scribe. https://www.logitech.com/en-gb/products/video-conferencing/room-solutions/scribe.html. Accessed 20 Nov 2023
3. Microsoft Teams content detection system. https://support.microsoft.com/en-au/office/share-whiteboards-and-documents-using-your-camera-in-microsoft-teams-meetings-905b52e3-bcd7-45c5-84cc-03992d7fc84f. Accessed 20 Nov 2023
4. Valueva, M.V., Nagornov, N.N., Lyakhov, P.A., Valuev, G.V., Chervyakov, N.I.: Application of the Residue Number System to Reduce Hardware Costs of the Convolutional Neural Network Implementation. Mathematics and Computers in Simulation, vol. 177, pp. 232–243. Elsevier BV., Amsterdam (2020)
5. Kota, B.U., Stone, A., Davila, K., Setlur, S., Govindaraju, V.: Automated whiteboard lecture video summarization by content region detection and representation. In: 2020 25th International Conference on Pattern Recognition (ICPR), Milan (2021)
6. Selinger, P.: Potrace: a polygon-based tracing algorithm, 20 September 2003. https://potrace.sourceforge.net/potrace.pdf. Accessed 20 Nov 2023

Studies at a Specialized University in the Post-pandemic Period from the Students' Point of View: New Realities and Challenges

Edita Butrime[1]($\boxtimes$) and Virginija Tuomaite[2]

[1] Lithuanian University of Health Sciences, A. Mickevicius St. 9, 44307 Kaunas, Lithuania
edita.butrime@lsmu.lt
[2] Kaunas University of Technology, K. Donelaičio St. 73, 44249 Kaunas, Lithuania

Abstract. Studying at university in the post-pandemic period is an interesting and challenging process that requires adaptation and innovation. Despite the difficulties, these new realities also provide opportunities to create more efficient, flexible, and diverse study programs. In this article, we will look at how university studies are changing and how students and universities are adapting to post-pandemic realities. Thus, the aim of this paper is to reveal students' attitudes towards studies in the post-pandemic period.

Keywords: distance learning · virtual learning during the post-pandemic · virtual learning environment of university

1 Introduction

The pandemic that started in 2019 and had a global impact on society also required universities to adapt to new realities. From distance learning to safety protocols, these changes have left a lasting mark on the university's community. In the spring of 2020, universities paid special attention to the transfer of teaching/learning activities to the distance teaching/learning, i.e., 'emergency remote teaching and learning' [1]. Under such conditions, it was impossible to perform a complex analysis of what tools to choose and how to organize the study process in the most rational way. In this article, we will look at how university studies are changing, and how students and universities are adapting to post-pandemic realities.

Thus, the aim of this paper is to reveal students' attitudes towards studies in the post-pandemic period.

In this paper e-learning is analysed as a socio-cultural system [2–5]. Such an approach towards e-learning enabled the presentation of 'a multi-dimensionality' in the concept. The analysis of e-learning as a socio-cultural system enabled the formulation of an interdisciplinary problem, for the solutions of which it is necessary to invoke theories and outcomes of computer science, culture, and education. The analysis allowed the enumeration of the forms and contents related to educational support for the participants (lecturers and students) of the system.

Á. Rocha et al. (Eds.): WorldCIST 2024, LNNS 988, pp. 142–152, 2024.
https://doi.org/10.1007/978-3-031-60224-5_16

2   Studies in the Post-pandemic Era

Quite a few scientific sources provide various predictions for the post-pandemic period. Hargreaves [6] believes that temporary distance teaching/learning will return to the traditional campus model or a hybrid approach. Meanwhile, Rashid and Yadav [7] argue that distance teaching/learning should remain an essential part of future studies, even when universities return to traditional (face-to-face) studies. In addition, states and universities are looking for creative solutions in the post-pandemic era. For example, in Sri Lanka, education policy makers are encouraging educators to gain more knowledge and skills in distance teaching/learning and assessment methods, and learners to demonstrate responsibility, organization, and commitment to the benefits of distance learning. However, researchers found that education policy makers need to allocate sufficient funds and improve distance education infrastructure [8]. Another situation is in the United Kingdom, where universities, after 2 years of fighting the pandemic, began to observe the reverse action, i.e., the transition from remote (distance) to traditional face-to-face teaching. Students are encouraged to study in the traditional way and that all students return to campus by a certain date. The conducted research showed that both internal and external challenges to students during the online-to-offline shift, which lead to a general resistance to the above-mentioned shift. The reduction of digital tools and learning materials in the VLE (virtual learning environment) was also a challenge for students who relied on digital resources for distance teaching/learning. Other challenges have also been identified, including academic barriers and problems with social engagement [9].

The results of a study conducted by Lamanauskas and Makarskaite-Petkeviciene [10] at one of the largest Lithuanian universities showed that first-cycle students usually evaluate distance studies positively, because this study format is convenient (the most important aspects are accessibility, flexibility, and compatibility with work activities). Another important advantage is the cost-effectiveness of studies (saving time and money). When applying distance studies in the post-pandemic period, the quality of studies deteriorates due to the lack of various effective methods used in distance studies. Unexploited opportunities can be seen in improving the presentation of educational content, increasing interactivity, ensuring better feedback, paying more attention to students' independent work, promoting communication in student groups. Administration of study processes is not a problem area. Despite the identified shortcomings, the possibilities of continuing studies remotely in the post-pandemic period are evaluated extremely positively (dominant evaluation). This concerns two significant issues: the appropriateness and usefulness of such studies. The negative evaluation of distance learning after the end of the pandemic is insignificant [10].

A student study [11] conducted in several countries (Lithuania, Japan, Sweden, and the UK) revealed that virtual learning environments, synchronous learning, and flipped classrooms are the dominant learning strategies that engage learners. The findings also showed that distance teaching/learning challenges stem from inadequate ICT infrastructure and digital literacy, health-related impairments, and professional and personal commitments that lead to learning interruptions. Thus, the study concluded that the overall social and vulnerability context of learners should be taken into account in order to increase the inclusion of distance teaching/learning.

A study conducted at a multi-campus university in South Africa [12] revealed a lack of learning tools, increased workload for teachers, problems with applied technology, and the need to monitor students' mental health problems. Other challenges arising from large student populations, high data processing and internet connection costs, and frequent power outages have been cited as consequences of existing social development gaps in South Africa. In the post-pandemic period, Ifenanyi [12] recommends the following methods of solving challenges: there is a need to monitor student mental wellness in the post-pandemic because some students are still traumatized by the menace of the pandemic; subsequent studies should assess and monitor the teaching and learning challenges students face post-pandemically; consultations and discussions to encourage the ongoing digitization are relevant and must be encouraged; in solving the problems of infrastructure development, ICT and e-learning, greater cooperation of the University management and relevant stakeholders is necessary; social development needs to be improved in most strategic locations in South Africa.

In conclusion, it can be said that the studies discussed here highlight different post-pandemic experiences, which can be named as subculture-influenced aspects.

3 Post-pandemic Studies in Higher Education Institutions of Health Sciences

In specialized higher education institutions such as the University of Health Sciences, the study programs (Medicine, Odontology, Nursing, etc.) have a lot of practical work or work in a simulation center. Such specificity poses special challenges for the study process. Thus, the pandemic period was a serious challenge for such higher education institutions [13–15].

In the post-pandemic period, these higher education institutions must make decisions to return to traditional studies (face-to-face) or to adapt the lessons of the pandemic period and change study processes. Italian universities [16] apply the following methods: persisting in the solutions adopted during the pandemic by maintaining remote lessons [17]; re-establishing the pre-existing modes of educational delivery (in the so-called 'returning, back to normal' approach); shaping a 'new normality' according to the lessons learned during the pandemic [18–20]. Nursing education has been dramatically transformed and shaped by changing constraints over time; restoring pre-pandemic educational models while failing to critically assess the transformations implemented may sacrifice the extraordinary learning opportunity provided by this extreme period of COVID-19. Bassi, et. al. [16] state that the post-pandemic model of nursing education in Italy should be based on the 'new normality', considering the lessons of the pandemic and, on the one hand, preventing a return to the pre-pandemic routine and consolidating the changes implemented during the emergency. A multi-dimensional set of recommendations was created, forming a strategic action map, where the main message is the need to rethink the entire process of nursing studies, embodying digitization.

Samuel et al. [21] conducted a study that reflects a significant aspect of distance teaching/learning in the Distance Learning Lab activities in School of Medicine distance learning. The study revealed that distance learning affects two main stakeholders: faculty members (in this case, teachers, and study support staff), and students. Therefore, in the post-pandemic period, to successfully transition to distance learning, the needs of both groups must be considered. Initiatives and support strategies should recognize that medical students and faculty members are adults. Therefore, the theory of andragogy can help in the development of support strategies. It cannot be assumed a priori that younger teachers and medical students, as 'digital natives' [22, 23], will have the technological skills necessary for distance teaching/learning. Samuel et al. [21] recommend based on their research and experience the following:

1. When making recommendations to students, it is advisable to include study service personnel as facilitators. This ensures that students are informed of the expectations of the lecturers, and the study staff can also be informed of the concerns of the students.
2. Providing support resources in a variety of ways ensures that more faculty members and students are served in a way that is most appropriate for them.
3. Continuously conduct needs assessment. Technology changes quickly and situations can change unexpectedly, as the COVID-19 pandemic shows. A constant understanding of the needs of faculty and students ensures that the institution always provides the most appropriate assistance.

According to Samuel et al. [21] some distance teaching/learning strategies used during the pandemic will continue. Students have grown accustomed to the flexibility of recorded lectures and expect faculty and study staff to continue to provide them with this. Faculty members had the opportunity to engage expert guest lecturers from around the world to enhance their studies. In the post-pandemic period, the possibility of using distance learning is great, because it has already been tested. In this scenario, it is useful to have support units (e.g.: Distance Learning Lab), which are identified by the authors as a successful solution for studying during a pandemic. Staff in these departments can recognize and meet the unique needs of medical faculty members and students. Another activity of such a unit should be education so that faculty members and students know how to use the available technologies to improve the students' learning experience [21].

Vallo Hult et al. [24] analyzed residency studies in the post-pandemic period. The authors distinguished the following design aspects of virtual learning environments (VLE) in health professional training institutions:

1. To activate VLE participants, it is advisable to use mobile technologies. Teachers can plan study activities that encourage students to walk and talk through group assignments.
2. Consider the cognitive load in digital courses, which are provided in the VLE. It is advisable to adjust the course design, format and content to avoid fatigue and cognitive workload of sitting in front of a screen all day.
3. Use interactive tools. This helps to avoid the passivity of students, which can occur during synchronous video conferences (lectures). A variety of tools can be used to engage students, e.g. using surveys and quizzes related to the lecture topic.
4. The digital format increased participant engagement compared to traditional lectures. Therefore, teachers should focus on teaching. A VLE moderator or study assistant should help manage the digital platform and participant questions. In this way, the teacher's technical workload is reduced, and they can concentrate on lecturing without interruption.

According to Mincu [25], micro-changes occur smoothly at any time, but for the transformation to occur, we must build on the accumulated wisdom and potential implicit in system and university leadership. Finally, the difficulty of learning from others and comparison should not be considered insurmountable. The author claims that there are no universal solutions when changing the system (in this case the study system) because it can be the case that, in certain conditions, we borrow not only solutions but the problems they address, in the way these are rhetorically framed. Mincu [25] argues that the greater influence of the political, social, and economic context on educational institutions and leaders around the world should not be underestimated. However, universal solutions must be thought out and adapted to specific educational institutions. Thus, studies of changes in study processes in the post-pandemic period in individual educational organizations are relevant to clarify the local context in order to apply universal solutions to it.

In conclusion, it can be said that in the post-pandemic period, the usefulness of distance learning solutions tested during the pandemic can be re-checked. This allows you to choose the most useful distance teaching/learning solutions and apply them in the future. Useful solutions for distance studies in health higher education institutions according to the authors [16, 17, 21, 24, 25] are the following: hybrid (or mixed) studies, as the 'new normality'; rethinking (revision) of the study process in order to embody digitization - to apply distance studies when it allows to create added values in studies; application of interactive tools in distance studies; application of mobile technologies in distance studies; providing technical support to teachers, study administrators and students in a rapidly changing virtual learning environment; development and support of professional IT solutions for studies and their administration in order to individualize studies. It is necessary to consider the context of the higher education institution, which means that there are challenges to be overcome when applying standard IT solutions in distance teaching/learning.

4   Studies at a Specialized University in the Post-pandemic Period from the Students' Point of View

The study was conducted in the fall semester of the 2023/2024 study year. The research was carried out in the following three stages: 1) analysis of university normative acts; 2) analysis of class schedules; 3) qualitative research - content analysis of students' answers to an open question. During the third phase of the study, students worked independently in the virtual learning environment Moodle. There was a new interactive tool during the homework, i.e., the Moodle activity Lesson. The topic was 'University as a Virtual Organization'. Learning activities that students had to do were the following: reading, writing, watching short videos, and answering self-analysis questions. If the student answers the self-analysis question incorrectly, the system directs them to the part of the lesson where they can familiarize themselves with the same lesson content again. When the student successfully finishes analyzing the content of the lesson, he is presented with the last open question: How do you feel during distance studies? Thus, this paper presents an analysis of students' answers to the above question (25% of students said they prefer distance learning, 25% of students said they prefer hybrid studies, 50% of students said they prefer traditional studies).

Stage 1. Normative acts of the university. In the fall semester of 2023/2024, by order of the university rector, teachers were recommended to work traditionally, i.e., face to face.

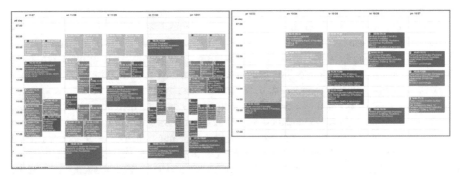

Fig. 1. Dentistry study program schedule (Left). Food Science study program schedule (Right). Remote classes are marked with red dots.

Stage 2. The one-week class schedule of the Dentistry study program was revised (Fig. 1). There are a total of 25 classes, 14 of which are distance learning. The one-week class schedule of the Food Science study program was revised. There are a total of 17 sessions, 4 of which are distance learning. On Thursday 2 remote and 1 traditional, on Friday 2 remote and 1 traditional.

Stage 3. Students who studied in the study programs of Dentistry and Food Science and were invited to complete the Study Introduction activities remotely. Those students who themselves requested that the course be conducted remotely were selected. The most frequently cited reason for the request was: '…inconvenient schedule…'. The number of students was 5 (Study program in Dentistry), and 8 (Study program in Food Science). These students were invited to do group remote work (the work was done in groups of 3 or 4 students), i.e., to draw a mind map 'The University of Distance Studies' (Fig. 2).

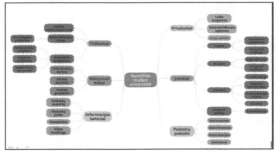
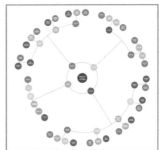

Fig. 2. Examples of mind maps.

The students were explained what a mind map is and how it is drawn. A list with links to online mind mapping tools was also provided. After the explanation, the students independently watched a video about creating mind maps. The group of students consulted among themselves and could choose the mind mapping tool that was the most attractive to them (Fig. 2).

The analysis is based on the concepts of the mind maps in question and the data extracted from them in categories (Table 1). The advantages of distance studies mentioned by the students were the following: saving time, recording lectures, saving university resources, convenience. The problems indicated by the students were as follows: lack of attention, lack of cooperation, technical disturbances, accessibility (expensive equipment, lack of training space), dishonesty. The behavior of participants in the distance learning process was described as follows: teachers are in a hurry, do not always communicate; students behave dishonestly, ignore other students or teachers (do not turn on cameras during synchronous remote communication, do not answer the questions of the teachers). The activities of the participants in the distance learning process were presented in the following way: teachers record lectures, prepare interactive tasks; students delve into the study material and summarize. The tools used were the following computer, microphone, tablet, video camera, headphones, and telephone.

Table 1. Students about their online study experiences in the post-pandemic period (analysis of the mind maps)

Categories	Subcategories	Subcategory assessment
Information sources	Lecture slides, lecture recordings, notes, professional video material	3 positive
Work with the teacher	Synchronous lectures, asynchronous lectures, consultations, seminars	positive
Methods of independent learning	Individual work, work in pairs, work in groups	positive
Communication	Social skills are lost (friends are lost, stress when talking to people)	negative
	Anxiety, depression	negative
	Lack of contact learning, communication with teachers (poorer learning outcomes, possibility to drop out of university)	negative

(continued)

Table 1. (*continued*)

Categories	Subcategories	Subcategory assessment
	Lack of discussions	negative
	Promotes cultural integration processes	positive
Flexibility	Dependence on connection and electricity sources	negative
	Workplace at home	negative
	Learning at their own convenience (time): Time management, Possibility to listen to recordings	positive
	Learning at your convenience (space): Comfortability	positive
Studying/learning	Less academic integrity	negative
	It is more difficult to concentrate: 'no participation' in the lecture	negative
	Lack of teacher's attention to the student	negative
	It is more difficult to adapt to information technologies	negative
	Improving computer literacy	positive
	Easier access to information (More knowledge, better learning results, all information in one place)	positive
	Development of competences	positive
Finance and resources	Use of electricity	negative
	Purchase of a computer and the necessary programs	negative
	Time saving	positive
	No need to travel by car or public transport (saves money and reduces pollution)	positive
	The university saves money: larger groups of students save energy, water, and other resources	positive

(*continued*)

Table 1. (*continued*)

Categories	Subcategories	Subcategory assessment
Tools	You need to purchase a phone, headphones, video camera, microphone, tablet, and computer	negative
	There is a phone, headphones, video camera, microphone, tablet, and computer	positive
Behaviour	Teachers are in a hurry	negative
	Teachers are not skilled in using remote communication tools, creating lecture recordings and interactive tasks	negative
	Sudents ignore remote classes	negative
	Sudents do not take notes	negative
	No being late	positive
	Tolerance	positive
	Mutual help	positive

Table 1 shows the categories and subcategories that were identified based on the analysis of the three student mind maps. There were 18 positive and 17 negative subcategories from the students' point of view. In conclusion, it can be said that students pay attention to the advantages and disadvantages of distance education. They are critical of these studies. The study revealed that students evaluate distance learning more positively than negatively. It is recommended that schedules are carefully made so that there are no remote and classroom sessions on the same day, as this is not convenient for students.

5   Conclusions

Studying at university in the post-pandemic period is an interesting and challenging process that requires adaptation and innovation. Despite the difficulties, these new realities also provide opportunities to create more efficient, flexible, and diverse study programs. This can determine the future generation's good direction and readiness for a rapidly changing global society.

In conclusion, it can be said that the studies discussed here highlight different post-pandemic experiences, which can be named as subculture-influenced aspects.

Useful solutions for distance studies in health higher education institutions according to the authors [16, 17, 21, 24, 25] are the following: hybrid (or mixed) studies, as the 'new normality'; rethinking (revision) of the study process in order to embody digitization - to apply distance studies when it allows to create added values in studies; application of interactive tools in distance studies; application of mobile technologies in distance studies; providing technical support to teachers, study administrators and students in a rapidly changing virtual learning environment; development and support of professional IT solutions for studies and their administration in order to individualize studies. It is necessary to consider the context of the higher education institution, which means that there are challenges to be overcome when applying standard IT solutions in distance teaching/learning.

References

1. Hodges, C., Moore, S., Lockee, B., Trust, T., Bond, A.: The difference between emergency remote teaching and online learning. EDUCAUSE (2020). Review
2. Scedrovitskij, G.P.: Izbrannye trudy. Skola kulturnoj politiki, Moskva (1995)
3. Mamardasvili M.K.: Procesy analiza i sinteza. Voprosy filosofii, No. 2. Moskva (1958)
4. Kvedaravicius, J.: Organizacijų vystymosi vadyba vadovelis/Kaunas (2006)
5. Butrime, E., Zuzeviciute, V.: E-learning as a Socio-cultural System: Towards a Balanced Development. LAP LAMBERT Academic Publishing, Saarbrücken (2014)
6. Hargreaves, A.: What's next for schools after coronavirus? Here are 5 big issues and opportunities. (2020). https://theconversation.com/whats-next-for-schools-after-coronavirus-here-are-5-big-issues-and-opportunities-135004
7. Rashid, S., Yadav, S.S.: Impact of Covid-19 pandemic on higher education and research. Indian J. Hum. Dev. **14**, 340–343 (2020). https://doi.org/10.1177/0973703020946700
8. Henadirage, A., Nuwan, G.: Retaining remote teaching and assessment methods in accounting education: drivers and challenges in the post-pandemic era. Int. J. Manag. Educ. **21**(2) (2023)
9. Zhao, X., Xue, W.: From online to offline education in the post-pandemic era: challenges encountered by international students at British universities. Front. Psychol. **13**, 1093475 (2023)
10. Lamanauskas, V., Makarskaite-Petkeviciene, R.: Distance education quality: first-cycle university students' position. Contemp. Educ. Technol. **15**(3) (2023)
11. Senanayake, A.C., et al.: Towards an inclusive disaster education: the state of online disaster education from the learner's perspective. Sustainability **15**(14), 11042 (2023)
12. Ifeanyi, F.O.: Barriers to learning linger into post-pandemic for multi-campus institutions in developing nations: a case of the University of the Free State. Soc. Sci. Human. Open **7**(1), 100438 (2023)
13. Butrime, E.: Virtual learning environments and learning change in modern higher education during the Covid-19 coronavirus pandemic: attitudes of university teachers. In: Rocha, Á., Adeli, H., Dzemyda, G., Moreira, F., Ramalho Correia, A.M. (eds.) WorldCIST. AISC, vol. 1367, pp. 222–231. Springer, Cham (2021). https://doi.org/10.1007/978-3-030-72660-7_22
14. Butrimė, E., Tuomaitė, V.: Temporary emergency and learning change in modern higher education during the COVID-19 pandemic: the attitudes of nursing students. In: Rutkauskiene, D. (ed.) Advance Learning Technologies and Applications. ALTA'20: Short Learning Programmes. Time for Learning is Now: Conference Proceedings, 2 December 2020. Kauno technologijos universitetas, Kaunas (2020)
15. Ahmed, H., Allaf, M., Elghazaly, H.: COVID-19 and medical education. Lancet. Infect. Dis. **20**(7), 777–778 (2020)
16. Bassi, E., et al.: Moving forward the Italian nursing education into the post-pandemic era: findings from a national qualitative research study. BMC Med. Educ. **23**(1), 452 (2023)

17. Silence, C., Rice, S.M., Pollock, S., Lubov, J.E., Oyesiku, L.O., Ganeshram, S., et al.: Life after lockdown: zooming out on perceptions in the post-videoconferencing era. Int. J. Womens Dermatol. **7**(5 Part B), 774–779 (2021)
18. Andraous, F., Amin, G.E.A., Allam, M.F.: The "new normal" for medical education during and post-COVID-19. Educ. Health (Abingdon) **35**(2), 67–68 (2022). https://doi.org/10.4103/efh.efh_412_20
19. Chu, L.F., Kurup, V.: Graduate medical education in anaesthesiology and COVID-19: lessons learned from a global pandemic. Curr. Opin. Anaesthesiol. **34**(6), 726–734 (2021). https://doi.org/10.1097/ACO.0000000000001065
20. Krohn, K.M., Sundberg, M.A., Quadri, N.S., Stauffer, W.M., Dhawan, A., Pogemiller, H., et al.: Global health education during the COVID-19 pandemic: challenges, adaptations, and lessons learned. Am. J. Trop. Med. Hyg. **105**(6), 1463–1467 (2021). https://doi.org/10.4269/ajtmh.21-0773
21. Samuel, A., Teng, Y., Soh, M.Y., King, B., Cervero, R.M., Durning, S.J.: Supporting the transition to distance education during the pandemic and beyond. Mil. Med. **188**(Supplement_2), 75–80 (2023)
22. Prensky, M.: Digital natives, digital immigrants. Horizon **9**(5), 1–6 (2001). http://www.scribd.com/doc/9799/Prensky-Digital-Natives-Digital-Immigrants-Part1
23. Prensky, M.: Digital natives, digital immigrants, part 2: do they really think differently? Horizon **9**(6), 1–6 (2001)
24. Vallo Hult, H., Master Östlund, C., Pålsson, P., Jood, K.: Designing for digital transformation of residency education-a post-pandemic pedagogical response. BMC Med. Educ. **23**(1), 1–10 (2023)
25. Mincu, M.: Why is school leadership key to transforming education? Structural and cultural assumptions for quality education in diverse contexts. Prospects **52**(3–4), 231–242 (2022)

Participating Elementary School Children in UI Design Process of Learning Environment: Case KidNet

Mirva Tapola[(✉)] [iD], Tuomas Mäkilä [iD], Norbert Erdmann [iD], and Mirjamaija Mikkilä-Erdmann [iD]

University of Turku, 20014 Turku, Finland
mikatap@utu.fi

Abstract. User interfaces are a significant part of websites, but even when websites are targeted at children, the UIs are designed by adults. Adults do not remember what it is like to be a child, yet children have usually been excluded from the design process, even when they are an important user group. The main goal of this study was to find a way to include 6th graders in the design process so that they could express what kind of UI elements they like. This study was conducted by organizing short design sessions where children got to choose their favorite UI elements from different design alternatives and use them to build a low-tech prototype of a UI. We found that the results produced with this method were sensible and revealed valuable information about children's UI preferences.

Keywords: Participatory Design · Child-Computer Interaction · Design Methods · User Interfaces

1 Introduction

This study investigated how primary school students perceive the user interface of a digital learning environment. In this study, the target group was fifth and sixth graders. Context for the study was a digital learning environment which was designed for learning Internet literacy skills in an everyday classroom situation. With Internet literacy skills, we mean searching for information on the internet, evaluating the reliability and usefulness of information, selecting the main ideas from the assessed as useful sources, and writing a synthesis to complete the inquiry task.

Participatory research design was used to engage children to contribute to the design process and find out their preferences about the UI design. In this study it is suggested that child development impacts the way children can use apps or webpages. Development has usually been divided into cognitive and physical development concerning especially motor skills [1–3]. Skills depend on the age and can differ from 1-year-olds non-existent to 14-year-olds adult-like skills. The need for help from the UI decreases as children's skills develop to the level of adults.

Children who participated in this study were mostly twelve and capable of similar physical interactions as the adults [4]. Cognitively they were at the beginning of the last of Piaget's four stages of development, the formal operations stage, where they learn to think systematically on abstract and hypothetical level [5].

The goal of this study was to enable children's participation in the design process and find out their preferences about the UI design of the learning environment. To achieve this goal the following research questions are analyzed in this study:

RQ1. How can user interfaces be designed with children?
RQ2. What kind of role can children have in the UI design?
RQ3. What kind of user interface would children design for KidNet?
RQ4. What kind of UI elements address the target group?

The article is organized as follows. Section 2 goes through the related research about children as users and participating children in the design process. The research approach is introduced in Sect. 3 where the case study description, protocol and implementation are presented. The results of the study are introduced in Sect. 4. The results are discussed in Sect. 5 and the conclusion and future work in Sect. 6.

2 Related Research

Children are a big user group with a lot of variances in skills needed to use a computer. UIs designed for children should follow the same basic principles of good UI design as the UIs for adults and therefore many of the same design recommendations created for adults are same for children's UIs [4, 6]. The fact that children and adults have different skills and goals creates a need for different UIs [7]. Adults use computers usually for productivity while the goals that children have are focused mostly on play and learning [1, 3, 4]. Adults usually know how to read and, although large amounts of text in the UIs should generally be avoided [8], the use of text in the UIs of child users should be minimized. This is because children's reading skills depend on the age and can vary from non-existent to good. Consequently, the used font should have good legibility and the font size should be larger especially for children who are beginning to learn how to read [1–4]. Children's developing motor skills should also be considered in the UI design. Young children need larger targets and simpler interactions, e.g. scrolling and dragging and dropping items can be difficult for young children [3, 4, 9].

When participating children in the UI design process, they can have various roles. Already in 2001 Druin [10] recognized that children could have four different roles in the design of new technology: user, tester, informant, and design partner. Two roles have been later proposed: child as a co-researcher and child as a protagonist [11]. During the last decade there have been many articles written about children participating in the design process. The way children take part in the design process depends on the role. Children have been included in one-time design sessions as informants where they have been asked to draw UIs and warning messages [12, 13]. Children have also been included in the design process as design partners, where they have been involved in the design process from the beginning to the end [14, 15]. The methods used depend on the age of the children involved in the design process. Younger children took part by answering questions and drawing [14]. Older children can take part in more complex ways like brainstorming, prototyping, and testing virtual reality intervention [15].

3   Research Approach

3.1   Case Study Description

Children grow up surrounded by technology, but they still lack the skills to distinguish advertisements from the real content [4] and have issues evaluating the reliability of websites [16]. KidNet is a closed online inquiry learning environment where children can practice Internet literacy. With Internet literacy skills we mean searching for information on the internet, evaluating the information, selecting the main ideas from the assessed as useful sources, and writing a synthesis to complete the task [17, 18].

Internet literacy is an essential part of the curricula in the Nordic schools. However, the digitalization in schools seems still to be a challenge. For example, international studies like the 2018 International Computer and Information Literacy Study (ICILS) indicate that every third Finnish student has not on reached the expected proficiency level. There are many other studies that prove students seem to have difficulties finding relevant www-sources and use these skills for learning purposes [19, 20]. Furthermore, Internet literacies are often not systematically trained in the Nordic countries [21, 22].

Children have been included in various roles throughout the design process of Kid-Net. Before the development started, children had the role of a user as they used the predecessor of KidNet called Neurone in a research project [23]. The data gathered then was analyzed and used in the design. Children also had the role of a tester in an early development phase. Finally, children had the role of an informant during these design sessions.

This case study uses the UI of KidNet as a context for the study. The UI for KidNet had to be suitable for children in elementary school. Because KidNet is used for learning new things, it had to be simple and easily learned so that children could focus on learning skills needed to complete the tasks given to them, instead of learning how to use KidNet. This highlights the importance of good usability and therefore KidNet UI provides a solid foundation for this study.

To keep the study setting feasible only one sub-page of the KidNet application was selected for the participatory study. The page chosen was the pageview-page presented in Fig. 1. In this page the analyzed web content is displayed. The users can save the page, evaluate the usefulness and reliability of the page, and save main points from the text to be used later when writing the answer to the assignment. In addition, KidNet has assignment, search, pageview, main points, bookmarks, and synthesis pages, but these have simpler functionality than the pageview-page and are therefore less suitable for participatory study.

3.2   Case Study Protocol

Research questions were addressed with a physical "UI jigsaw puzzle" that had different UI elements (e.g., navigation bar, buttons, etc.) with multiple designs for each element. A UI design tool called Figma [24] was used to create different UI elements for the study. The elements were designed with usability in mind, but some of them deviated from the most traditional web UI components so that children could choose something different if they wanted to. The final elements are illustrated in Fig. 2.

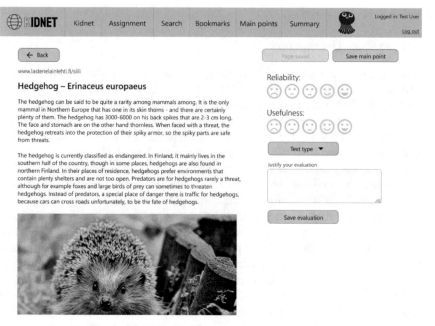

Fig. 1. KidNet's pageview-page in December 2022

The focus of the design sessions was to find out which UI components the participants liked the most and what kind of UI they would design for KidNet. The study group had used KidNet before and were told that the current UI was designed by adults who do not remember what children like and that they now have a chance to make changes to the UI. The control group had not used KidNet before and were asked to design a UI with the components they liked the most. The importance of children's opinions was highlighted by telling them that they are the main users of KidNet and therefore their opinions really matter.

Design sessions began with a brief introduction about the session's purpose. The participants were then asked a couple of background questions involving their access to a computer and their own estimate about their computer skills. The participants were then asked to design a UI that has a navigation bar, an image area, a text area, buttons for key functions, evaluation buttons and a support button. The participants chose their favorite components and created a layout with the chosen components. When the layouts were ready, the participants were asked to pick colors for different elements. The design sessions ended with questions regarding decorations, how the support element should function and if children thought the task was fun. The sessions were recorded with an audio recorder to be transcribed and the final UI designs were photographed to be recreated with Figma. The choices of each participant were entered into an Excel sheet for further analysis.

Fig. 2. UI elements used in the design sessions. Navigation: text-only, icons-only, icons with text and balls with text. Buttons: Pill-shaped, rounded corners, sharp corners, a circle, and a square. Fonts: Arial, Times, and Comic Sans. Evaluation: emojis, stars, Likert scale and school grading. Support characters: question mark, octopus, robot and KidNet hero.

3.3   Case Study Implementation

Thirteen children (10–12 years; 6 girls, 7 boys) participated in the design sessions. Participants can be divided into two groups: 1) the study group of six 6th grade students (3 girls, 3 boys) from Rauma normal school who had previously used KidNet during the pre- and post-tests of a parallel research and 2) the control group of seven children (4th grade: 1 girl, 2 boys; 6th grade: 2 girls, 2 boys) from Turku area who had not previously used KidNet. The control group was used to validate the results of study group and see how the use of KidNet impacted the choices children made.

As discussed previously in Sect. 1, children who participated in this study were getting closer to adult users in the sense that their fine motor skills are developed enough, so that there is no need for larger buttons to support the poor fine motor skills, they know how to read, so they do not rely on images to understand the meaning of different buttons and their cognitive skills are getting closer to adults level, meaning they are capable of solving more complex tasks than younger children. [4, 5].

A test session was held at the University of Turku's Educarium building in December 2022 with a 4th grader. The purpose of the session was to test if the components and the study protocol worked, but it was also included in the control group. The design sessions with the study group were held at the Rauma normal school in December 2022. Participants came to a small room behind the actual classroom after they had finished the post-test of the parallel KidNet research.

The control group sessions were held during spring 2023 at the University of Turku and at the homes of the participants. The control group had a bigger age distribution (4th and 6th graders), and the sessions were arranged in different environments than study groups' sessions because they were mostly held at the homes of the participants. One of the children who came to university got exposed to a poster on the wall of the room the session was held in. This might have impacted the choices he made. Three control group sessions were held at the homes of the participants. Two sessions with two children and one session with one child. Sessions with two children were arranged so that peaking or eavesdropping was impossible for the child waiting for their turn.

4 Results

4.1 Illustrations of UI Designs

In this section sample designs from both the study group and the control group are presented. The selected designs illustrate some of the key findings of the study.

The designs in Fig. 3 are from the study group, both were done by 6th grader girls. The design on the left has different elements compared to KidNet's UI, but the layout is identical to the layout used in KidNet at the time (Fig. 1). The design on the right stood out from the other designs in the study group. The choices about elements and colors and the placement of elements differed from the other designs.

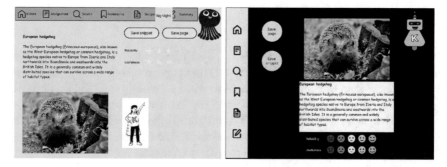

Fig. 3. Two designs from the study group

The design on the left of Fig. 4 was made by a 6th grader girl in the control group. It was one of the control group designs that had the navigation bar at the bottom of the screen. This participant was the only participant in the control group who chose Comic Sans for the font. The placement of the image and text differed from the others as well, most of the participants placed the image over the text.

The design on the right of Fig. 4 was made by a 4th grader boy in the control group. He was the only participant in the control group who chose Arial for the font but the choices for other elements were like other children's. However, there are enormous differences in the layout and the use of colors. Other layouts had more similarities and colors were used mostly on the background, navigation, and buttons.

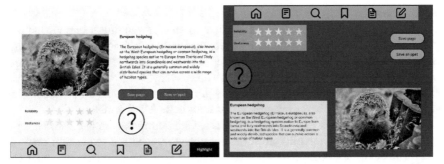

Fig. 4. Two designs from the control group

4.2 Comparison of UI Designs

Navigation

The choices children made about the style of navigation were similar in both groups. Children in study group were used to the text-only option as it was used in KidNet but still 67% of them chose the icons and text option and 33% of them chose the icons-only option. Children in the control group made similar choices (icons-only 29%, icons and text 71%). Text-only and balls were not chosen by anyone.

Buttons

The pill-shaped button was chosen by 60% of the children. The choices made by children in the two groups were similar. The button with rounded corners was chosen three times and the ball-shaped button was chosen only in the study group twice. The button with sharp corners and the square were not chosen by anyone.

Font

The choices children made about the fonts can be seen in Table 1. There was a large difference in the choices made by children in the study group and children in the control group. Comic Sans was chosen four times in the study group while Times was chosen five times in the control group. The fact that Arial was chosen in each group only once is interesting since sans-serif fonts are usually used in UIs because of their legibility in smaller sizes, so the font should have been a familiar UI font for children [7].

Table 1. Font choices

Font	Study group	Control group	Combined
Arial	1 (16,7%)	1 (14,3%)	2 (15,4%)
Comic	4 (66,67%)	1 (14,3%)	5 (38,5%)
Times	1 (16,7%)	5 (71,4%)	6 (46,2%)
Total	6 (100%)	7 (100%)	13 (100%)

Support character
Participants were asked "what would you like to click if you didn't know what to do on the page?". There was a significant difference in the choices made by children in the study group and children in control group (see Table 2). Most of the children in the control group grabbed the question mark as soon as they saw it without thinking about other options. However, nobody in the study group chose the question mark and chose the octopus instead, probably because it was used in the UI of KidNet.

In total, the octopus was chosen six times. Children in the study group chose the octopus four times while children in the control group chose the question mark five times. The octopus was chosen only twice in the control group, where the other child that chose it had been exposed to an octopus-poster on the wall.

Table 2. Support character

Support character	Study group	Control group	Combined
Question mark	0 (0%)	5 (71,4%)	5 (36,5%)
Octopus	4 (66,67%)	2 (28,6%)	6 (46,15%)
Girl	0 (0%)	0 (0%)	0 (0%)
Robot	2 (33,3%)	0 (0%)	2 (15,4%)
Total	6 (100%)	7 (100%)	13 (100%)

Evaluation
Stars were chosen seven times for evaluating the reliability and usefulness of the page. The choices made by children in the study group and children in the control group were different. Emojis were chosen three times in the study group and only once in the control group, the children in study group were used to the emojis since it was used in KidNet which explains the difference. The stars were chosen five times in the control group and twice in the study group. Likert and school grade were both chosen once. The school grade was chosen by a child in the control group. He reasoned that if KidNet is used in schools, then the school grading should be used.

Layout
The layouts had lots of variations as can be seen in Sect. 4.1. There were differences between the control group and study group, and it seems like the previous use of KidNet had an impact on the layouts. One of the children in the study group recreated the UI layout from KidNet, other children grouped certain elements together and placed them in similar positions to what they were in the KidNet UI. E.g., buttons and evaluation were grouped together and placed like they are in the actual UI. The layouts made by children in the control group did not have these similarities, there were more gaps and empty space, and the placement of the elements did not have similarities.

The most obvious difference was the placement of the navigation bar (Table 3). Children in the study group placed the navigation bar to the top or the left side while three in the control group placed it to the bottom. The reasons for this need to be studied more, but there is one known difference between children in study group and children in control group that might play a significant role in this case: children in study group were using laptops in school while children in control group were using tablets where tool bars placed in the bottom of the screen are commonly used.

Table 3. The position of navigation bar

Navigation's position	Study group	Control group	Combined
Top	2 (33,3%)	1 (14,3%)	3 (23,1%)
Left	4 (66,7%)	2 (28,6%)	6 (46,2%)
Bottom	0 (0%)	3 (42,9%)	3 (23,1%)
Right	0 (0%)	1 (14,3%)	1 (7,7%)
Total	6 (100%)	7 (100%)	13 100%)

5   Discussion

The results of this study had practical implications supporting the UI design process of KidNet. The results validated some of the design choices made for the actual KidNet's UI. The results also suggest that children like traditional and familiar UI elements more than unusual and unfamiliar UI elements.

The fact that children in study group chose the octopus character for support validates the design choice made previously. At the same time, using the octopus character should be critically analyzed because it became clear that children in the control group would be searching for a question mark in a situation where they need help. Children in the study group liked the octopus and it makes the UI funnier, but deviating from UI design norms might cause challenges with wider audience.

The popularity of Times in the control group is conflicting with previous research that specified that children would prefer Comic Sans as a font and Times should be avoided because it is considered to be less easy to read [25]. Serif fonts like Times are often used in printed material so the use of printed elements could impact this choice. Arial has been shown to be more legible than Times for all grades in another research [26], validating the use of sans-serif font in the UI.

It can be said that the method used in this study proved to be useful in finding out the preferences of children regarding the different designs of UI elements. The study showed that some of the layouts that children created had severe usability issues (see images in Sect. 4.1), this was however expected since designing UIs is a learned skill. Children seem to prefer familiar and usable UI elements instead of "childish" and unfamiliar elements. Children seemed to like the visual appearance of fun and cartoonish designs for some elements. In this case, using these elements in the UI proposes no threat to high usability, which is important in a learning environment. The main purpose of this study was to gain valuable information for professional UI designers about the preferences children had instead of having children create finished UIs.

It should be noted that because the elements were designed based on design recommendations the results are probably more useful than if the children would have been asked to draw the UIs by themselves. The method was efficient in the sense that the sessions lasted 6 min on average and the results can be objectively analyzed. The downside is that the creativity of children was limited because they e.g. did not have an opportunity to draw their own elements if they did not like the existing designs.

At the general level, it is likely that the participatory method can also be used in the development of user interfaces for other applications aimed at children. This article provides a very concrete method for involving children in the design process in the role of an informant. It is likely that the method could be useful also when children have the role of a design partner. Although the topic has been widely studied in the past, the methods used in the study were not found in the research literature.

6   Conclusion and Future Work

The goal of this study was to enable children's participation in the design process and find out their preferences about the UI design. Short design sessions were organized with two groups of fifth and sixth graders. During the sessions the children choose their preferred UI elements for an individual page of the KidNet application from a pre-made set of UI elements. Based on the empirical evidence gathered during the study and related research, the research questions of this study can be answered as follows:

RQ1. How can user interfaces be designed with children? There are various methods for designing UIs with children as discussed in Sect. 2. In this study children participated in the design process with the help of the low-tech prototype. The results suggest that the method provides useful information about children's preferences (**RQ1**). The method was efficient, and the results were easily analyzed but it came with the price of children's freedom to draw their own UIs.

RQ2. What kind of role can children have in the UI design? Children can have various roles in the design process as discussed in Sect. 2. In this study children had the role of informant (**RQ2**).

RQ3. What kind of user interface would children design for KidNet? Based on the results, children would create a UI with a navigation that has icons and text, pill-shaped buttons, and stars for evaluating the page (**RQ3**). The choices were adult-like in the sense that the chosen elements could be used in KidNet. The previous use of KidNet had an impact on the component choices and layouts that children made. This suggests that the use of KidNet impacted the children's mental model of a website.

RQ4. What kind of UI elements address the target group? Children made similar choices about navigation and buttons. However, the previous use of KidNet had an impact on the choices children made and there were differences especially concerning the support and evaluation elements.

Further research is needed to better understand children's UI element preferences. It would be beneficial to analyze whether UIs designed based on the children's preferences lead to better usability scores amongst the target group. Also, the participative method could be further investigated by analyzing how suitable the method is for different age groups and whether it could be used when children participate in the UI design process as design partners.

The study showed objective details about children's UI element preferences and proved the feasibility of the presented participative design method both as a practical tool for children's UI design and as a research instrument. It is important to have concrete methods and tools for participating children in the UI design process because adults do not remember what it is like to be a child.

References

1. Hourcade, J.P.: Child-Computer Interaction, 2nd edition. http://homepage.cs.uiowa.edu/~hourcade/book/hourcade_cci_2nd_edition.pdf. Accessed 21 Apr 2023
2. Jacko, J.A.: Human Computer Interaction Handbook: Fundamentals, Evolving Technologies, and Emerging Applications, Third Edition. Taylor & Francis Group, Baton Rouge, UNITED STATES (2012)
3. Chiasson, S., Gutwin, C.: Design Principles for Children's Technology. 9
4. Liu, F., Sherwin, K., Budiu, R.: UX Design for Children (Ages 3–12). Fremont, CA : Nielsen Norman Group (2022)
5. Crain, W.C.: Theories of Development: Concepts and Applications. Prentice Hall, Boston (1992)
6. Nielsen, J.: 10 Usability Heuristics for User Interface Design. https://www.nngroup.com/articles/ten-usability-heuristics/. Accessed 20 Nov 2023
7. Tidwell, J., Brewer, C., Valencia, A.: Designing Interfaces: Patterns for Effective Interaction Design. O'Reilly Media, Incorporated, Sebastopol, United States (2020)
8. Johnson, J.: Designing With the Mind in Mind: Simple Guide to Understanding User Interface Design Guidelines. Morgan Kaufmann (2020)
9. Hourcade, J.P.: Learning from preschool children's pointing sub-movements. In: Proceedings of the 2006 Conference on Interaction Design and Children, pp. 65–72. Association for Computing Machinery, New York, NY, USA (2006). https://doi.org/10.1145/1139073.1139093
10. Druin, A.: The role of children in the design of new technology. Behav. Inf. Technol. **21**, 1–25 (2001). https://doi.org/10.1080/01449290110108659
11. Iversen, O.S., Smith, R.C., Dindler, C.: Child as protagonist: expanding the role of children in participatory design. In: Proceedings of the 2017 Conference on Interaction Design and Children, pp. 27–37. Association for Computing Machinery, New York, NY, USA (2017). https://doi.org/10.1145/3078072.3079725
12. Kraleva, R.S.: Designing an interface for a mobile application based on children's opinion. Int. J. Interact. Mob. Technol. (iJIM). **11**, 53–70 (2017). https://doi.org/10.3991/ijim.v11i1.6099
13. Dempsey, J., Sim, G., Cassidy, B., Ta, V.-T.: Children designing privacy warnings: Informing a set of design guidelines. Int. J. Child-Comput. Interact. **31**, 100446 (2022). https://doi.org/10.1016/j.ijcci.2021.100446
14. Yip, J.C., Ello, F.M.T., Tsukiyama, F., Wairagade, A., Ahn, J.: Money shouldn't be money!: An examination of financial literacy and technology for children through co-design. In: Proceedings of the 22nd Annual ACM Interaction Design and Children Conference, pp. 82–93. Association for Computing Machinery, New York, NY, USA (2023). https://doi.org/10.1145/3585088.3589355
15. Kitson, A., Antle, A.N., Slovak, P.: Co-designing a virtual reality intervention for supporting cognitive reappraisal skills development with youth. In: Proceedings of the 22nd Annual ACM Interaction Design and Children Conference, pp. 14–26. ACM, Chicago IL USA (2023). https://doi.org/10.1145/3585088.3589381

16. Hämäläinen, E.K., Kiili, C., Marttunen, M., Räikkönen, E., González-Ibáñez, R., Leppänen, P.H.T.: Promoting sixth graders' credibility evaluation of Web pages: an intervention study. Comput. Hum. Behav. **110**, 106372 (2020). https://doi.org/10.1016/j.chb.2020.106372
17. Bauer, A.T., Mohseni Ahooei, E.: Rearticulating internet literacy. J. Cyberspace Stud. **2**, 29–53 (2018). https://doi.org/10.22059/jcss.2018.245833.1012
18. Livingstone, S.: Internet literacy: Young people's negotiation of new online opportunities. In: Matrizes, p. 11 (2011). https://doi.org/10.11606/issn.1982-8160.v4i2p11-42
19. Kiili, C., et al.: Reading to Learn from online information: modeling the factor structure. J. Lit. Res. **50**, 304–334 (2018). https://doi.org/10.1177/1086296X18784640
20. Hautala, J., Kiili, C., Kammerer, Y., Loberg, O., Hokkanen, S., Leppanen, P.: Sixth graders' evaluation strategies when reading Internet search results: an eye-tracking study. Behav. Inf. Technol. **37**(8), 761–773 (2018). https://doi.org/10.1080/0144929X.2018.1477992
21. Sormunen, E., et al.: A Performance-based Test for Assessing Students' Online Inquiry Competences in Schools. Presented at the September 18 (2017)
22. Sundin, O.: Invisible search: information literacy in the swedish curriculum for compulsory schools. Nord. J. Dig. Literacy **10**, 193–209 (2015). https://doi.org/10.18261/ISSN1891-943X-2015-04-01
23. González-Ibáñez, R., Gacitúa, D., Sormunen, E., Kiili, C.: NEURONE: oNlinE inqUiRy experimentatiON systEm. Proc. Assoc. Inf. Sci. Technol. **54**, 687–689 (2017). https://doi.org/10.1002/pra2.2017.14505401117
24. Figma: The Collaborative Interface Design Tool. https://www.figma.com/. Accessed 05 Jul 2023
25. Chaparro, B., Bernard, M., Mills, M., Frank, T., McKown, J.: Which Fonts Do Children Prefer to Read Online? 1 (2023)
26. Woods, R.J., Davis, K., Scharff, L.F.V.: Effects of Typeface and Font Size on Legibility for Children

Analyzing University Dropout Rates Using Bayesian Methods: A Case Study in University Level in Mexico

José Ramón de la Cruz[1], Diana L. Hernández-Romero[1], Pedro Arguijo[1] (ORCID),
Luis Carlos Sandoval Herazo[2] (ORCID), and Roberto Ángel Meléndez-Armenta[1(✉)] (ORCID)

[1] Division of Graduate Studies and Research, Tecnológico Nacional de México/Instituto Tecnológico Superior de Misantla, Km 1.8 Carretera a Loma del Cojolite, 93821 Misantla, Mexico

{182t0876,212t0437,sparguijoh,ramelendeza}@itsm.edu.mx

[2] Wetlands and Environmental Sustainability Laboratory, Division of Graduate Studies and Research, Tecnológico Nacional de México/Instituto Tecnológico de Misantla, Veracruz, Km 1.8 Carretera a Loma del Cojolite, 93821 Misantla, Mexico

lcsandovalh@itsm.edu.mx

Abstract. In Mexico's higher education, student dropout issue is critical, influenced by multiple factors. The Instituto Tecnológico Superior de Comalcalco (ITSC) recorded 3,499 students across eight engineering and two bachelor's programs in 2018, experiencing a significant dropout rate of 9%. This article presents a web platform using Bayesian methods to analyze and interpret dropout data from ITSC. Specifically, the Computer Systems Engineering program, most impacted by dropouts, was analyzed to identify underlying patterns and factors. Utilizing Naive Bayes models, the platform categorizes students and assesses dropout probabilities, offering insights for data-driven strategies to mitigate this challenge in higher education.

Keywords: University Dropout · Bayesian Approach · Decision Trees

1 Introduction

University education is an academic phase during students acquired specific technical, professional, human, and disciplinary skills to contribute to the economic, technological, and social growth. One of the most significant challenges faced by higher education institutions is student dropout, defined as time when any student decides to don't continue their studies before receiving their degree or diploma, and without having been transferred to another institution [1] or in student's perspective, associated term with concepts like departure, withdrawal, academic failure, non-persistence, and non-completion due to a long series of interactions between the student and the academic and social systems of educational institutions [2]. Several governmental organizations have own definitions of school dropout; for example, the Secretary of Public Education (SEP) in Mexico defines it as an indicator expressing the number or percentage of students who leave academic activities before completing a grade or educational level.

In Mexico, school dropout is an issue that affects different educational levels, being higher education the most impacted. The National Institute of Statistics and Geography (INEGI) in Mexico recorded that the average schooling of the population over 15 years of age continues to increase, but only 21% have university education studies such as a bachelor's or master's degree [3]. In 2018, INEGI recorded in its annual census that, out of every 100 university students, only 8 complete their studies, and only 38% of higher education students achieve the goal of graduating, as reported by [4].

The Tecnológico Nacional de México (TecNM) is one of the most relevant universities with 254 campuses across the country, including the Instituto Tecnológico Superior de Comalcalco (ITSC), located in the city of Comalcalco in the state of Tabasco, in the southeastern zone. This university has 3,499 students distributed across 8 engineering programs and two bachelor's degrees, but also has faced various problems with school dropout. According to information obtained from the historical data on student academic performance for the school years 2010–2018 at this university, an average school dropout rate of 9% was recorded in 2018 across the eight programs offered at this higher education institution.

In ITSC, the engineering career with the highest number of dropouts during this school period was Computer Systems Engineering (ISC), with 43 dropouts, corresponding to 14% of the overall dropout rate. This research presents a procedure to identify behavioral patterns among students of ISC in the ITSC and the relationship with factors influencing school dropout. Additionally, the research includes an analysis of the subjects and the gender of the student during their educational experiences using a Bayesian approach. In this context, the research question guiding this work is as follows:

- RQ1: What are the recurring patterns among dropout students in Computer Systems Engineering?
- RQ2: What is the difference between the gender of the students and their academic performance during their time in engineering?

The present research is organized as follows: Sect. 2 presents related works with the main topics of the current research. Then, in Sect. 3, showing the proposed methodology for this research. In the Sect. 4 the results of identifying recurring patterns among dropout students obtained through a Bayesian approach are presented. Finally, in Sect. 5 the conclusions of this research work are discussed, including works and applications to be developed in the future.

2   Related Works

In the last decade, school dropout rates have significantly increased, prompting educational authorities to explore new techniques to reduce the number of students leaving school before obtaining their degree. Understanding and relating the different factors present in dropout students is crucial. Recently, the use of data science algorithms as emerged as an alternative to identify behavioral patterns in higher education dropout students.

In this context, [5] presents a predicting model of risk of student dropout and to help improve pedagogical practices in students likely to drop out of school using five classification algorithms: 1) K-NN, which is an algorithm based on setting several points $k(n)$ that exist in a certain region centered on a vector of x features, by growing a region around said vector, with an appropriate metric, until k points are captured, these are the $k(n)$ nearest neighbors of x [9], 2) Naive Bayes, which is a classifier based on Bayes' theorem, which considers the variables to be predicted x as the root and the attribute variables x_i as leaves, furthermore, this classifier assumes conditional independence between features to classify new unobserved examples of the variable x after training with maximum likelihood [10]. 3) Decision Three (DT), which is an algorithm that takes as input an object or a situation described through a set of attributes and returns a decision, where the input attributes can be discrete or continuous [10]. 4) Random Forest (RF), which is an algorithm defined as the collection of another classifier, the decision tree and that can be constructed using a method called bagging along with random selection of attributes and providing a training set. In addition, random forests consider many fewer attributes for each division, they are fast to perform classification, and their accuracy is considerably good [11], and 5) Support Vector Machine (SVM), which is learning efficient used in a wide variety of classification and regression applications and is generally used to represent decisions rather than posterior probabilities according to groups having x attributes [10, 12]. The predictive model uses a cleaned and pre-processed data collection of 26 students with 19 attributes, including academic and personal data, was used. The results shows that the proposed methods were tested with 8 groups with different characteristics, being group number 6, the best performed with best across all algorithms, trained and tested with features such as obtained marks, scenario, higher studies, number of transfers, monthly income, and parents' occupation. In this group, the best classifiers were DT and SVM, which recorded a 100% accuracy rate.

On other hand, [6] propose a method based on machine learning using DT, C4.5, RF CART, Naive Bayes, and SVM techniques, with a dataset comprising student data about socio-academic information of first year of the 2010–2011 generation at the University of Oviedo, in Asturias, Spain. Starting with a sample population of n observations (n = 1055, in this sample), the data were classified into three categories: Dropout, Change, and Continue. Each model was then trained using default configurations and repeating the experiments 10 times per model. The results showed that the C4.5 method is slightly better than other methods, except SVM, and slightly worse than all methods, except CART, in terms of accuracy.

School dropout leads to a shortage of professionals in various fields in the labor market, and one way to reduce or completely eradicate this problem is to identify students at risk of dropping out. In [7], data mining techniques were used to build predictive models with RF, DT, Naive Bayes, Logistic Regression (LR), which is a model to deduce the logarithmic probabilities of class membership using discriminants and linear regression classifiers [13] and KNN algorithms to identify students who may drop out in the first year of a Computer Science career from the University of the South Pacific in Suva, Fiji. All models were trained and tested employing ten-fold cross-validation for three stages, aiming to evaluate accuracy. The results focus on two models, with Naive Bayes performing better in stage 1 with an AUC of 0.6132, but in stages 2 and 3, the overall performance of the LR models was better with an AUC of 0.7523 and 0.8902, respectively.

When dealing with a considerably large amount of data, it is necessary to apply data mining techniques to manage and process such information. In [8], the management of a data warehouse is demonstrated by applying four models based on Machine Learning (ML) algorithms: DT, KNN, LR, and Naive Bayes classifier, with the aim of extracting relevant information and predicting university student dropout during their first three years at university. The student attributes included demographic background, geographic origin, socioeconomic index, academic performance, university performance, financial indicators, among others. The DT and KNN models used 21 variables to evaluate their performance, and the LR and Naive Bayes models used 41 variables respectively. In the results section, it is shown that it is possible to predict student dropout with an accuracy that exceeded 80% in most scenarios across the different models applied.

3 Methodology

3.1 Dataset

The ITSC stores student information in an academic tracking system, a web platform unique to the institution, where the academic progress of students. The acquisition of this information allowed for the generation of a dataset with the following characteristics (variables) of the students: student ID, year of entry, semester, status, number of credits in the curriculum, gender, subject, credits, semester of the curriculum, current semester, grade, and accreditation modality.

Table 1. Scholar data information of university students of ITSC.

Variable	Type	Encoding
Student ID	String	[TE100035, TE130921]
Year of entry	Nominal	{2010, 2011, 2012, 2013}
Semester	Nominal	{1, 2, 3, 4, 5, 6, 7, 8, 9, 10, 11, 12, 13}
Status	Nominal	{graduates, dropouts, degree}
Number of credits in the curriculum	Numeric	260
Gender	Nominal	{male, female}
Subject	Nominal	{differential calculus, discrete mathematics, ethics workshop, etc.}
Credits	Nominal	{4, 5, 6, 10}
Semester of the curriculum	Nominal	{1, 2, 3, 4, 5, 6, 7, 8, 9}
Current semester	Nominal	{1, 2, 3, 4, 5, 6, 7, 8, 9, 10, 11, 12, 13}
Grade	Nominal	{0, 70, 71, 72, 73, 74, 75, 76, 77, 78, 79, 80, 81, 82, 83, 84, 85, 86, 87, 88, 89, 90, 91, 92, 93, 94, 95, 96 97, 98, 99, 100}
Accreditation modality	Nominal	{first evaluation, second evaluation, first repeat, second repeat, special}

The information is stored in a file in comma-separated values format, with a CSV extension, which is the starting point for the analysis and identification of patterns associated with school dropout in engineering students. The description of each of the variables that make up the dataset is shown in Table 1.

3.2 Procedure

Figure 1 shows the procedure proposed in this research, divided into two main stages. The first, pattern recognition, involves the acquisition, analysis, and processing of data to retain only the most valuable information. The cleaned and processed data are used in a model based on Bayesian approach. In the second stage, decision making, the developed model is integrated into an intelligent web platform to identify recurrent patterns in school dropout, generating useful conclusions and information for teachers and educational university authorities.

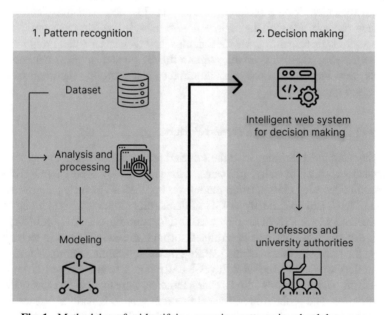

Fig. 1. Methodology for identifying recurring patterns in school dropouts.

3.3 Data Processing and Transformation

The preparation of the acquired data was carried out through cleaning and selection, to obtain consistent information, eliminate irrelevant values, and achieve better results in classification algorithms. This process was divided into three stages: 1) Extraction, where variables provide substantial data to identify recurrent patterns in students who abandon the ISC program were extracted, 2) Transformation of data through necessary modifications were made to the selected variables, and labels were assigned to the new attributes, all through coding in the Python programming language, and 3) Storage, where the processed data are saved in a spreadsheet files in XLSX and CSV formats.

To understand student behavior in relation to school dropout, it was necessary to analyze data concerning their grades and failed subjects, across three different categories: Graduates, Degree Holders, and Dropouts. According to the analysis of the obtained data, subjects in the basic engineering sciences area are most associated with school dropout in the first three semesters of university. Specifically, Differential Calculus is one of the subjects with the highest failure rate; 77% of dropout students in their first semester of university did not pass this subject. In the second semester of the 71% of students who dropped out during this period, 17% failed this subject. Lastly, during the third semester, 61% of dropouts did not pass differential calculus.

Furthermore, this study emphasizes the gender variable, where the corresponding attributes are male and female. Of the dropout students in the dataset, 278 are male and 84 are female, translating to 76.8% of total dropouts being male and 23.2% female. From the male perspective, the five subjects that led to an increase in dropout numbers are: Operations Research, Data Structures, Vector Calculus, and Probability and Statistics. For females, the subjects with the highest failure rates, influencing dropout, are Operations Research, Data Structures, Vector Calculus, and Object-Oriented Programming. In the data processing stage, the variables of enrollment, gender, subject, semester, grade, and status were identified as the most important for discovering recurrent patterns in dropout based on grades.

3.4 Model to Identify School Dropout Patterns

To applying statistical techniques to the selected and processed data to predict scholar status students within university: graduates, degree holders and dropouts, a Naive Bayes-based machine learning classification model was applied. To identify patterns in school dropout in ITSC, and specifically in ISC students, this model was used along with the dataset obtained where 75% of the data was used for training, and 25% for testing. Then, the quality of the classifier was determined through cross-validation, a technique for evaluating the accuracy of a machine learning model by training multiple models.

For the development of the Naive Bayes-based model, all records from the knowledge base, consisting of 526 records with 10 characteristics, were loaded. To classify the target value, restriction, and probabilities, Bayes' theorem was used, and three types of Naive Bayes predictors were implemented. Results are in Table 2; the summary is the following:

- Categorical Naive Bayes: This algorithm had an estimated accuracy of 80% given the training dataset introduced into the model.
- Gaussian Naive Bayes: The input data for the model had to be converted and represented using a Bernoulli distribution for the characteristics, and the target variable was represented with a labeling of [0,1,2]. The accuracy obtained with this algorithm was 80%.
- Multinomial Naive Bayes: The model developed with this algorithm concluded with favorable figures, showing an estimated accuracy of 80%, as it used a weighted average by support.

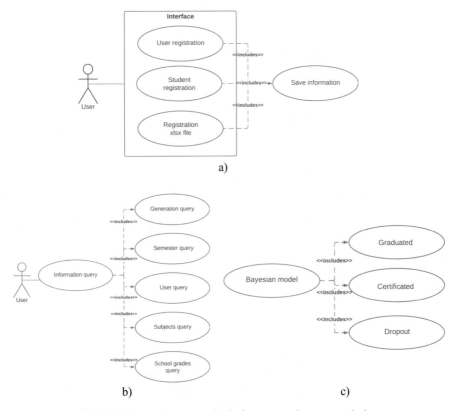

Fig. 2. Software dropout web platform general use cases design

3.5 School Dropout Web Platform

To process new data and show the interpretation of the developed model of student dropout, a web platform was developed that shows and record the following information:

- Schoolar students' information (Fig. 2.a)
- Charts and diagrams that provide general information for decision-making (Fig. 2.b)
- Information and results for decision-making model approach (Fig. 2.c).

4 Results

To gain a broader perspective on which gender of students tend to drop out during university studies, the model was tested with new data collection divided from men and women. For the implementation of the model with the male gender knowledge base, 398 collected data records with their 10 corresponding characteristics were used. The results of the three algorithms with the Naive Bayes model performance through cross-validation are showing in Table 3 Similarly, the Bayesian model was tested with the knowledge base containing data from female students, consisting of 127 records with their 10 corresponding characteristics. The results of the three algorithms with the Naive Bayes model performance through cross-validation are showing in Table 4.

Table 2. Comparison of the Naive Bayes models with the original data.

Model	Correctly classified instances	Correctly classified instances	Accuracy
Categorical Naive Bayes	105	27	80%
Gaussian Naive Bayes	105	27	80%
Multinomial Naive Bayes	105	27	80%

Table 3. Comparison of Naive Bayes models with data from male students.

Model	Correctly classified instances	Correctly classified instances	Accuracy
Categorical Naive Bayes	87	13	86%
Gaussian Naive Bayes	80	20	80%
Multinomial Naive Bayes	86	14	86%

Table 4. Comparison of Naive Bayes models with data from female students.

Model	Correctly classified instances	Correctly classified instances	Accuracy
Categorical Naive Bayes	29	3	91%
Gaussian Naive Bayes	29	3	91%
Multinomial Naive Bayes	29	3	91%

Regarding the interface, as a result, a web platform was obtained where clean and preloaded data provided by ITSC can be consulted on the page, and new data can be added to the system's database.

It is possible to display the prediction of the Naive Bayes-based model to classify students as graduates, degree holders, or dropouts, according to whether they passed the corresponding subjects or not, as shown in Fig. 3.

Finally, it is possible to consult a prediction by subject and gender, with the aim of obtaining more specific conclusions about the number of male and female students who drop out, graduate, or earn degrees, and what techniques could be applied to provide educational support to the indicated population.

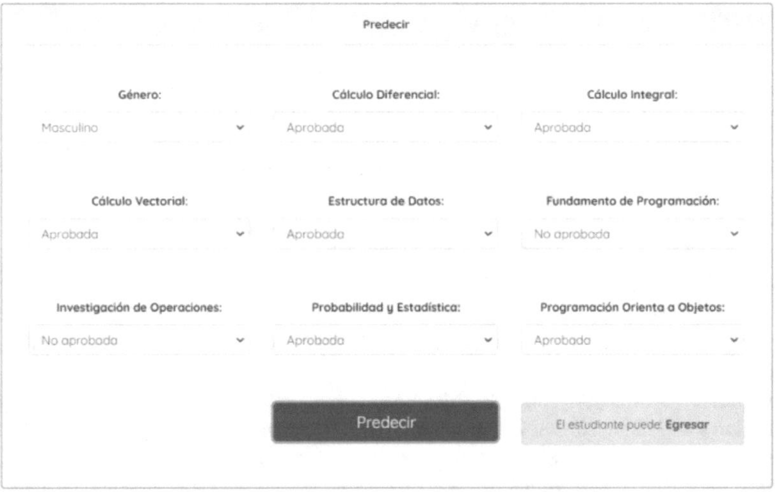

Fig. 3. General student prediction interface

5 Discussion and Conclusion

The intelligent web platform developed for this study can classify a selected dataset of students from ITSC (Instituto Tecnológico Superior de Comalcalco) from the 2010 to 2018 academic years and predicting whether they will be dropouts, graduates, or degree holders, using algorithms based on a Bayesian approach. This work contributes to the field of study by utilizing data mining techniques and overcoming the limitation of not being able to analyze and use large datasets. This research focuses solely on understanding the academic behavior of ITSC students, with an interest in observing how students' academic abilities develop based on their grades in each subject taught throughout ISC program.

According to the results in the analysis stage, feature reduction was carried out to identify patterns and characteristics associated with school dropout. However, the research delved deeper into the causes of school dropout according to the gender of the student. The results and analysis of this study confirm that male students have different academic performance abilities compared to female students. Additionally, tests on the intelligent web platform show that men have difficulty passing specialty subjects. On the other hand, it is evident that women lack the ability to pass basic science theories.

Finally, it is essential to consider some ethical implications and considerations in the use of predictive models to identify potential cases of school dropout. For example: 1) consent, ensuring the protection of personal data and explicit permission for its use; 2) equity, avoiding models that reinforce pre-existing inequalities; 3) transparency and ability, ensuring that the processes and decisions of the model are clear and understandable; and 4) supervision, establishing control mechanisms and continuous evaluation to ensure the proper application and effectiveness of these models.

References

1. Rochin Berumen, F.L.: Deserción escolar en la educación superior en México: revisión de literatura. RIDE. Revista Iberoamericana para la Investigación y el Desarrollo Educativo **11**(22) (2021)
2. Kim, D., Kim, S.: Sustainable education: analyzing the determinants of university student dropout by nonlinear panel data models. Sustainability **10**, 954–972 (2018)
3. Otero Escobar, A.D.: Deserción escolar en estudiantes universitarios: estudio de caso del área económico-administrativa. RIDE. Revista Iberoamericana para la Investigación y el Desarrollo Educativo **12**(23) (2021)
4. Ricárdez, D.H., Olán, L.V., Leija, H.G.: Causales de la deserción escolar en México, perspectiva desde un análisis documental. TECTZAPIC: Revista Académico-Científica **7**(3), pp. 9–20 (2021)
5. Jaiswal, G., Sharma, A., Sarup, R.: Machine learning in higher education: Predicting student attrition status using educational data mining. In: Handbook of Research on Emerging Trends and Applications of Machine Learning, pp. 27–46 (2020). IGI Global
6. Rodríguez-Muñiz, L.J., Bernardo, A.B., Esteban, M., Díaz, I.: Dropout and transfer paths: what are the risky profiles when analyzing university persistence with machine learning techniques? Plos one **14**(6) (2019)
7. Naseem, M., Chaudhary, K., Sharma, B.: Predicting freshmen attrition in computing science using data mining. Educ. Inf. Technol. **27**, 9587–9617 (2022). https://doi.org/10.1007/s10639-022-11018-3
8. Palacios, C.A., Reyes-Suárez, J.A., Bearzotti, L.A., Leiva, V., Marchant, C.: Knowledge discovery for higher education student retention based on data mining: machine learning algorithms and case study in Chile. Entropy **23**(4), 485 (2021). https://doi.org/10.3390/e23040485
9. De Sa, J.M.: Pattern recognition: concepts, methods, and applications. Springer Science & Business Media (2001). https://doi.org/10.1007/978-3-642-56651-6
10. Russell, S., Norvig, P.: Artificial Intelligence: A Modern Approach. Pearson Higher Education, Saddle River (2019)
11. Han, J., Pei, J., Tong: Data mining: concepts and techniques. Morgan Kaufmann (2012)
12. Bishop, C.: Pattern Recognition and Machine Learning. Springer (2006). https://doi.org/10.1007/978-0-387-45528-0
13. Provost, F., Fawcett, T.: Data Science for Business: What you need to know about data mining and data-analytic thinking. O'Reilly Media Inc, Sebastopol (2013)

Cloud Computing Platforms to Improve Shuar Chicham Writing Skills

Pablo Alejandro Quezada-Sarmiento[1](✉) , Patricia Marisol Chango-Cañaveral[2] ,
Janio Jadán-Guerrero[3] , and Washikiat Pedro Tsere-Juma[4]

[1] Centro de Investigación de Ciencias Humanas y de La Educación-CICHE, Facultad de
Ciencias de La Educación, Universidad Indoamérica, Bolivar 2035 y Guayaquil, 180103
Ambato, Tungurahua, Ecuador
pabloquezada@uti.edu.ec

[2] Universidad Técnica Particular de Loja- Facultad de Ciencias Económicas y Empresariales,
1101608 Loja, Ecuador
pmchango@utpl.edu.ec

[3] Centro de Investigación en Mecatrónica y Sistemas Interactivos – MIST, Universidad
Indoamérica, Av. Machala y Sabanilla, Cotocollao, Quito, Ecuador
janiojadan@uti.edu.ec

[4] Universidad Intercultural de Las Nacionalidades y Pueblos Indígenas Amawtay Wasi, Carrera
de Lengua y Cultura, Av. Colón 5-56 y Juan León Mera, Edif. Ave María, Torre B, Pichincha,
Ecuador
washikiat.tsere@uaw.edu.ec

Abstract. Higher education through virtual environments is centered on the student, oriented towards interactive learning, in situations that are close to the real world, for which teachers require new communicative, non-verbal skills and an innovative approach to learning, which will accompany their students. Students in the complex process of acquiring even more knowledge of a non-native language. In the same context, the use of open-source technology platforms is becoming the best way to deliver solutions that meet the current need for the process of acquiring an ancestral language, so in this document we will focus on the analysis and use of Cloud Computing Tools as an alternative to complementary material to improve writing skills in the process of Second Language Acquisition (SLA) in our case Shuar Chicham.

Likewise, teachers should not focus only on online learning but also on interaction with students as Gibson (1993) states; thus, innovative technologies must be means that help reduce the digital technological gap, allowing solutions, answers and agile questions to be given within the teaching-learning process, which is visible with the use of cloud computing tools.

Therefore, using cloud computing tools it was proposed to demonstrate if writing skills can be improved in students who learned Shuar Chicham basic.

Keywords: Ancestral language; Cloud Computing · Education; Technology · Platforms · Language · Shuar Chicham · Writing Skills

Á. Rocha et al. (Eds.): WorldCIST 2024, LNNS 988, pp. 175–183, 2024.
https://doi.org/10.1007/978-3-031-60224-5_19

1 Introduction

Shuar Chicham.is an integral part of the cultural heritage of Indigenous communities in Ecuador specialty of the Amazon Region. By learning Shuar Chicham, individuals contribute to the preservation and revitalization of this language, which is crucial for maintaining cultural identity and passing down traditional knowledge and values to future generations. Learning Shuar Chicham enables effective communication with Shuar-speaking individuals, fostering understanding and connection between different communities. It allows for meaningful interactions, exchange of ideas, and the building of relationships based on mutual respect and cultural appreciation.The use of recent technologies and programs can be used in various fields, including education, which will serve for the design, search, presentation, exchange, and reuse of material because the technology allows storing, organizing, replicating, disseminating, transform and be accessible, which leads to saving time and resources [1].

The educational field has seen the need to create more spaces where there are no limitations of time or space; Thus, distance education uses various platforms for the teaching-learning process.

According to [2] the use of open-source platforms generates greater interaction, encourages teamwork among students and develops positive feedback between teachers and students. Under this premise, this document proposes the use of open-source technology platforms to improve writing skills in the acquisition of a Second language in our context Shuar Chicham.

2 Context

Ecuador is a multicultural and multiethnic country that has 14 nationalities and 19 peoples, where the original nationalities are settled, and the language of th,ese nationalities must be strengthened. Well, the Constitution of Ecuador itself in its Art. 2 makes official that:" Spanish is the official language of Ecuador; "Spanish, Kichwa and Shuar are official languages of intercultural relations." In the same way, the Model of the MOSEIB Intercultural Bilingual Education System, in its Art. 3, requires all teachers of the Bilingual Intercultural System to use the Shuar language in all comprehensive training processes from Early Childhood Family Community Education to Baccalaureate. With the above, it shows how important it is to learn the Shuar language to have a job, and not only in the educational field, but also in other public positions. If we learn an ancestral language from our country, we value and strengthen the original languages to provide continuity from generation to generation.

The use of multimedia resources, the Internet, digital communications, and electronic resources have spread widely, especially as aids to learning. Computers and electronic communication tools have played a significant role in education because they allow teachers to develop interactive classes. Thus, computer and electronic tools help the development of education as they allow the creation of more effective and productive learning environments where teachers and students share knowledge and learn in more practical and innovative classes. At present, the interest in using them as tools to help education has grown the most.

For this reason, [3] states that technology has always been implicitly central to the understanding of what Shuar Chicham. is or can be. After all, without manuscripts and pens and without paper and printers, technologies would not be texts or anything to read, and with no historical dimension to the subject matter itself. This process was the basis for developing innovative technologies, especially in the teaching of Shuar Chicham. In addition, innovative technologies, and the wide world of the web (Internet) offer a wide range of opportunities for both teachers and students interested in improving Shuar Chicham language skills.

[4] states that chat rooms, online forums, and discussion groups provide practice writing in Shuar Chicham, with the opportunity to get feedback on grammar and spelling. These innovative technologies are rapidly changing the face of education. In this changing educational environment, it is essential to improve the principles and design a process that indicates how to satisfy students with effective educational experiences.

2.1 Collaborative Learning

Collaborative learning is an effective approach to teaching and can be particularly valuable when working with diverse groups, such as the Shuar people. The Shuar are an Indigenous people of the Amazon rainforest, primarily located in Ecuador and Peru. When designing a collaborative learning environment to teach the Shuar, it's essential to consider their cultural context, language, and educational needs.

[5] mentions that collaborative work is a relationship between students that requires positive interdependence (a feeling of sinking or swimming together), individual responsibility (each of us has to contribute and learn), interpersonal skills (communication, truth, leadership, making decisions and conflict resolution), promoting face-to-face interaction, and processing (reflecting on how well the team is working and how it could work better). Students learn best when they are actively involved in the process. Researchers mention that regardless of the subject matter, students who work in small groups tend to learn more of what is taught and retain longer than the same content that is presented in other educational formats. Students who work in collaborative groups also appear more satisfied with their classes.

[6] states that formal learning groups are teams established to complete a specific task, such as developing a laboratory experiment, drafting a report, developing a project, or preparing a document. These groups can complete their work in a single class or in several weeks.

Typically, students work together until the assignment is complete, and their project is graded. In this research, the formal learning groups are Shuar Chicham level I.

Collaborative learning for teaching the Shuar involves a holistic and culturally sensitive approach. By incorporating their language, culture, and community into the educational process, you can create a more meaningful and effective learning experience for Shuar students.

2.2 E – Writing

Teaching the Shuar Chicham through written materials involves creating resources that are culturally sensitive, linguistically appropriate, and engaging.

[7], mentions that E-writing is writing done without the medium of the computerized network. Technology has an important impact on the learning and development of writing skills; for example, computer technology, especially word processing, has become a vital writing tool for students. Increased students are looking to improve their technology skills through activities and experiences made available through technology especially with the use of electronic tools.

The integration of open-source technology platforms, specifically Cloud Computing Tools, is highlighted as a key solution to address the current needs of foreign language acquisition, specifically in the context of learning Shuar Chicham as a non-native language. The document suggests that these tools can serve as complementary materials to enhance writing skills in the Second Language Acquisition (SLA) process.

The key is to create e-learning materials that are not only educational but also respectful of the Shuar culture and community. Regularly seek feedback from the Shuar learners and community members to ensure that the materials remain relevant and effective.

2.3 Alfresco

It is an enterprise content platform that can be used in the cloud or behind a security system. It helps to store and share several types of documents and resources in business management. In the educational field, it has great applicability in the creation of repositories and control of application projects that simulate a real environment.

In the Fig. 1 the Alfresco platform is show.

Fig. 1. Alfresco interface used in this research.

2.4   Redbooth

It is another of the cloud tools used, also known as TeamBOX, which is a collaboration and communication platform that provides a place for shared tasks, discussions, file sharing, group chat and video conferences. It is flexible and easy to use, allowing project teams and teachers to conduct their work in a professional and interactive way [5].

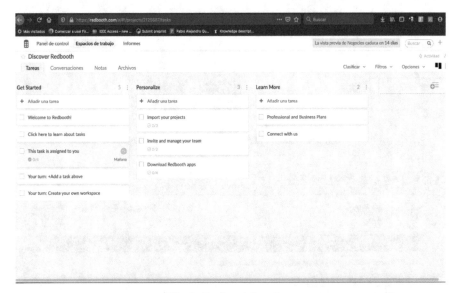

Fig. 2. Redbooth interface.

The Redbooth cloud tool is like having a social network to manage projects, like a Facebook for the office. It is a solution to improve productivity, that is, its main feature is to allow the creation of projects with different people and to observe and share information directly with the project participants, uploading files, analyzing the time spent on each task (Fig. 2).

For the process of analysis of the development of writing skills in the Kichwa I language, the open-source platforms were used: Alfresco and Redbooth.

The participation of a teacher from the Ancestral Language of "Universidad Intercultural de las Nacionalidades y Pueblos Indígenas Amawtay Wasi" (UINPIAW) thirty students of the Shuar Chicham level I component of education were participants of this study. In the Fig. 3 the Moodle platform used is showed.

In the same context gamification method were used to improve the writing skills. In the Fig. 4 a game to improvement writing skills is present.

Using the Alfresco platform, sites were created where each group of students planned and created the different activities regarding the components. Additionally, through this tool it was possible to generate the respective Feedback and the generation of knowledge related to the component.

Fig. 3. Moodle platform Shuar Chicham class

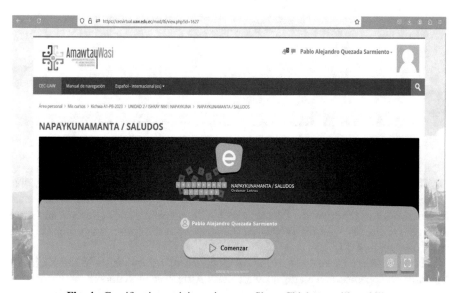

Fig. 4. Gamification activity to improve Shuar Chicham writing skills.

Likewise, the Redbooth tool was used for feedback between teacher-students and student-students of the various activities developed within the educational component. To select the tools, the evaluation criteria indicated in Table 1 are required.

In the table below the main characteristics of Alfresco and Redbooth are shown.

Table 1. Cloud Tools Evaluation Criteria adapted of [7]

Criteria	Alfresco	Redbooth
Share, manage tasks	Yes	Yes
Collaborative component	Yes	Yes
Integration with cloud tools	Yes	Yes
Notifications via email	Yes	Yes
User Management	Yes	Yes
Accessibility	Yes	Yes
Payment License	Yes	Yes

3   Results

The results obtained in the development of this research are detailed below:

By learning Shuar Chicham, students can access and learn from this traditional ecological knowledge, contributing to the promotion of sustainable development, environmental stewardship, and the preservation of traditional practices that are harmonious with nature. In the Fig. 5 the video about culture and traditions are showed. With this resource the student learns about Shuar history and make a brief essay about this video, where the writing skills were improvement.

Fig. 5. Video activity to improvement the writing skills

Fig. 6. Shuar Chicham writing classes.

Greater interactivity between teacher and students(yachakuk) and student-students (yachakuk-yachakuk). In the Fig. 6, Shuar writing class is showed.

Access to information on each of the activities conducted. Linking of accounts, accesses, and notifications according to each tool.

Generation of digital portfolios and Feedback.

The key is to integrate these cloud computing tools into a comprehensive language learning strategy that includes speaking, listening, reading, and writing exercises. Additionally, engaging with native speakers and immersing oneself in the language and culture is crucial for a holistic learning experience.

4   Conclusions

By learning Shuar Chicham, individuals contribute to creating a more inclusive society where different languages and cultures are valued. It promotes linguistic diversity and challenges linguistic discrimination, helping to overcome language barriers and promoting social integration.

Learning Shuar Chicham holds immense importance in terms of cultural preservation, fostering communication and understanding, promoting inclusivity, expanding opportunities, personal growth, and contributing to sustainable development. It is a way to honor and respect the cultural diversity and knowledge systems of Indigenous communities in Ecuador.

The use of open-source platforms: Alfresco and Redbooth allowed us to generate an environment of greater interactivity between teachers and students.

Generating collaborative networks through open-source platforms allowed greater interaction between students and student-students, which generated a collaborative social environment.

The development of this research allowed us to generate Feedback and consider improvements in the context of teaching and learning within the teaching of a second language.

At the end of the Shuar-Chicham A 1 course, we can realize that the Shuar language is very important for the teachers and administrators of the Amawtay Wasi University; since the mission and vision of the university is "to form human beings who recognize the harmonious relationship between all beings of life, who fully exercise their individual and collective rights for the construction of the Plurinational and Intercultural State, based on good "community living" and aims to deconsolidate by promoting intercultural dialogue aimed at Good Living (Tarimiat Pujustin). This Shuar Chicham A1 learning course promotes intercultural relationships between teachers, administrators, and students, where respect, reciprocity and strengthening of the Shuar language is generated. We are sure that Amawtay Wasi University will promote and be the cradle of ancestral knowledge of the peoples and nationalities of our country.

Proficiency in Shuar Chicham opens new opportunities for employment and engagement with Shuar Chicham -speaking communities, particularly in fields such as education, healthcare, tourism, and community development. It allows individuals to work more effectively in multicultural contexts and to engage in cross-cultural collaborations.

References

1. AREA, M.: La educación en el laberinto tecnológico. De la escritura a las máquinas digitales. Barcelona, OctaedroEUB (2005)
2. Quezada-Sarmiento, P.A., Suárez-Guerrero, C., Narvaez-Rios, M.M., Gutiérrez-Albán, L.G.: Use of cloud computing tools on pedagogical and educational contexts (2022). https://doi.org/10.1007/978-3-031-04826-5_35
3. Pope, R.: An introduction to language, literature, and culture. London and New York: Rutledge Group (2007)
4. Nutt, A.: Learning English Using Technology (2006). http://www.eslteachersboard.com/cgi bin/english/index.pl?read=2064
5. Schoenbaum, D.: About Redbooth (2014). Disponible en: https://redbooth.com/about.
6. Srinivas, H.: Collaborative Learning (2010). http://www.gdrc.org/kmgmt/c-learn/index.html
7. Quezada-Sarmiento, P.A., Suárez–guerrero, C.: Cloud computing in the training process in web programming. [La computación en la nube en el proceso formativo en programación web] RISTI - Revista Iberica De Sistemas e Tecnologias De Informacao (E42), 10–19 (2021)

Beyond the Screen: Exploring Students' Views on Social Media's Impact in Education

Wegene Demeke[(✉)]

School of Computing, Engineering and the Built Environment, Edinburgh Napier University,
Edinburgh EH10 5DT, Scotland, UK
w.demeke@napier.ac.uk

Abstract. This study delved into the intricate relationship between social media usage and academic outcomes among university students from diverse fields: engineering and art studies, health and social care studies, and business studies. The study employed a two-pronged methodological approach, encompassing a survey questionnaire and focus group discussions. The survey data, reported in a separate publication, provided quantitative insights into students' social media usage patterns, preferences, and perceived impacts on academic performance. The focus group discussions, conducted across three university campuses, facilitated a more in-depth exploration of students' experiences, motivations, and the nuances of their social media usage behaviors.

The study's findings highlighted the complex interplay between social media and academic outcomes. While excessive social media engagement could negatively impact academic performance, social media could also be harnessed for collaboration, knowledge sharing, and peer-to-peer learning. Students' preferences for specific social media platforms were influenced by the purpose of use and the desire to maintain a sense of autonomy. The study concluded that educators should adopt strategies to help students manage their social media usage effectively, maximizing its potential benefits while minimizing distractions and negative impacts on academic performance.

Keywords: Social media · academic outcome · University students

1 Introduction

In the dynamic landscape of contemporary higher education, understanding the impact of social media usage on student attention, motivation, and self-regulation is of paramount importance. As social media platforms become increasingly intertwined with students' daily lives, their influence on academic engagement and performance cannot be overlooked. This research offers valuable insights into the complex interplay between social media usage and academic outcomes, providing a foundation for educators and policymakers to develop effective interventions to support students' success in the digital age.

University students use social media to interact with families and friends. This research is exploring on the use of social media of university students to prepare for their assessment and on how it also affects students negatively. The literature indicates the advantages of using social media by students for collaborative learning (Al-Adwan et al., 2020; Al-rahmi, 2013). Through the collaboration of social media, the work by (W. M. Al-Rahmi et al., 2014) suggested that social media use improves academic performance. However, there is a research gap on finding on students' perception on the effect of the use of social media on their preparation for assessment for how social media affects students' engagement and preparation for their assessment.

The following research questions are the drivers for this research. The rationale for the development of research question 1 (see Table 1) is that social media has become an integral part of many students' lives, and its use can have both positive and negative impacts on their academic performance. Understanding how social media use affects engagement, motivation, and self-regulation can help educators design interventions to maximise the benefits and minimise the potential drawbacks of social media use for academic purposes.

The second research question is developed to understand collaboration and group work. Social media can provide a platform for students to coordinate group projects, share ideas, and provide mutual support. However, the effectiveness of social media for collaboration may depend on factors such as the size and dynamics of the group, the clarity of communication channels, and the establishment of clear expectations and goals.

The final research question is about knowing what students' preference of social media. Understanding students' preferences for social media platforms for academic purposes can help educators and technology designers develop more effective and targeted social media-based learning interventions. By considering the factors influencing these preferences, such as academic discipline, learning style, and individual needs, educators can tailor their approaches to maximize engagement and learning outcomes.

Table 1. Research Questions

RQs	Description
RQ 1	How does the frequency and purpose of social media use influence students' academic attention, motivation influenced by self-regulation?
RQ 2	How does social media facilitate collaboration among students, including group project coordination, knowledge sharing, and peer-to-peer learning, and how does this collaboration impact students' academic outcomes?
RQ 3	What factors influence students' preferences for specific social media platforms for academic purposes?

The rest of the manuscript starts with a literature review that critically evaluates existing research on the topic, identifying key themes, gaps in the literature, and emerging trends. The methods section outlines the research design, data collection procedures, and analytical techniques employed in the study. The findings section presents the results of the study, highlighting significant patterns and relationships between social media use and academic outcomes. The discussion section interprets the findings in the context of existing literature and theoretical frameworks, offering explanations for the observed relationships and exploring potential implications for educators and policy-makers. Finally, the conclusion summarizes the key findings of the study, restates the significance of the research, and suggests directions for future research.

2 Literature Review

A vast majority of students are active on social media, as observed in the study by Kolhar et al., (2021), which found that 97% of students use these platforms. However, only a minor proportion, about 1%, utilize social media primarily for academic purposes. The study further revealed diverse usage patterns among students: 35% predominantly use it for chatting, 43% use it as a pastime, and a significant 57% exhibit high levels of engagement with social media, which may suggest a tendency towards habitual or excessive use. While the high usage of social media for non-academic purposes is evident, there are studies that reveal a different aspect of this trend.

Contrasting these findings, a study from South Africa by Chukwuere, (2021) suggested that social media use can positively influence students' academic learning performance. This improvement was noted in areas such as communication, interaction, engagement, and self-directed learning. A key reason for this positive impact is the access social media provides to academic resources. Platforms like YouTube offer an extensive range of educational content, including tutorials, lectures, and coursework from various institutions and educators globally. Such resources can significantly enhance understanding and supplement classroom learning. This positive influence is further exemplified in collaborative educational settings.

Social media has emerged as a powerful tool for fostering collaborative learning among students, facilitating interactivity with peers, and promoting knowledge sharing. These platforms provide a virtual space where students can connect, communicate, and engage in shared learning experiences that extend beyond the confines of the traditional classroom (Ansari & Khan, 2020).

Collaborative learning involves students working together to achieve common learning goals. Social media enables this by providing a platform for students to share ideas, brainstorm solutions, and co-create knowledge. Online discussion forums, group chats, and collaborative workspaces allow students to engage in asynchronous and synchronous discussions, fostering a sense of community and shared purpose. Interactivity with peers is another key aspect of social media's impact on student learning. Social media platforms provide opportunities for students to interact with their peers in a variety of ways, including asking questions, providing feedback, and offering support. This peer-to-peer interaction can enhance understanding, clarify concepts, and promote diverse perspectives, enriching the learning experience (W. Al-Rahmi et al., 2014).

Knowledge sharing, the exchange of information and expertise, is also facilitated by social media. Students can share study materials, post links to relevant resources, and engage in discussions that contribute to the collective knowledge base. This collaborative knowledge construction process deepens students' understanding and promotes the development of critical thinking skills. However, alongside these positive aspects, there are studies highlighting the challenges associated with social media use.

A parallel study by Bhandarkar et al., (2021) also highlighted the extensive use of social media among students. According to their findings, 71.5% of students utilized social media to complete assignments, and over 50% used it for preparing for seminars. The study noted a negative impact of social media on students' academic performance. Additionally, it was observed that academically lower-performing students tended to use social media more frequently compared to their higher-performing counterparts.

Social media usage has emerged as a prominent factor influencing student attention, motivation, and self-regulation, consequently impacting academic performance. The pervasiveness of social media platforms has created a constant stream of distractions, readily accessible at the touch of a button. This continuous influx of notifications, updates, and engaging content can significantly divert students' attention away from academic tasks, hindering their ability to focus and negatively affect their academic performance (Giunchiglia et al., 2018; Hassell & Sukalich, 2016).

Moreover, social media's ability to foster instant gratification and social connection can undermine intrinsic motivation, the inherent drive to engage in activities for personal satisfaction or interest. Instead, students may become more inclined to seek extrinsic rewards, such as the validation and approval of their peers, through social media engagement. This shift in motivation can diminish their genuine interest in academic pursuits, further affecting their attention and engagement (Barton et al., 2018; Thompson, 2017).

Self-regulation, the ability to manage one's thoughts, emotions, and behaviors, plays a pivotal role in mediating the effects of social media on attention and motivation. Students with strong self-regulation strategies are better equipped to resist distractions, manage their time effectively, and prioritize academic tasks. Conversely, students with weaker self-regulation skills may struggle to control their social media usage, leading to decreased attention, reduced motivation, and ultimately, impaired academic performance (Khan et al., 2021). Amidst these varied impacts, understanding student preferences in social media platforms remains a complex issue.

Despite the widespread use of social media among students, the existing literature does not provide a definitive understanding of the factors that influence their choices of platforms for specific purposes. While some studies have identified general trends, such as the use of platforms like Instagram and TikTok for entertainment and Facebook and WhatsApp for social interaction, there is a lack of clear consensus on the underlying factors driving these preferences.

This lack of clarity can be attributed to the complex and multifaceted nature of social media usage. Students may consider various factors when selecting platforms, including the platform's features, the type of content it offers, the perceived benefits of using the platform, and the individuals they wish to connect with. Additionally, individual preferences and motivations play a significant role in shaping social media usage patterns. Further research is needed to delve deeper into the factors that influence students' social media platform preferences. This research should explore the interplay between individual characteristics, social context, and technological affordances to shed light on the complex decision-making processes behind social media usage. By understanding these factors, educators and policymakers can develop more effective strategies to guide students towards mindful and purposeful social media engagement.

3 Methodology

To investigate the intricacies of students' social media usage and its influence on their academic preparation and outcomes, this study employed a two-pronged methodological approach. The first phase involved a survey questionnaire, the results of which have been reported in a separate publication. This survey provided valuable insights into students' social media usage patterns, preferences, and perceived impacts on their academic performance. Building upon the foundational data gathered from the survey, the study expanded its scope to include qualitative insights. Building upon the foundational data gathered from the survey, the study expanded its scope to include qualitative insights.

3.1 Rationale for Methodology Selection

The combination of a previously conducted survey and the focus group discussions provided a robust methodological framework for addressing the research questions. The survey offered a quantitative assessment of students' social media usage patterns and their perceived impacts on academic performance, laying the groundwork for the focus group discussions. To deepen our understanding of these initial findings, the study utilized a more interactive and detailed approach.

To further explore these findings and gain a deeper understanding of students' experiences, motivations, and the nuances of their social media usage, focus group discussions were conducted across three distinct university campuses. These discussions, encompassing students from engineering and art studies, health and social care studies, and business studies, facilitated a more in-depth examination of their perspectives and the contextual factors influencing their social media usage behaviors. To ensure the effectiveness and integrity of these discussions, a comprehensive data collection procedure was meticulously followed.

3.2 Data Collection Procedure

To ensure the quality and effectiveness of the focus group discussions, a thorough data collection procedure was implemented. This procedure encompassed the following steps:

1. Ethics Procedures: Ethical procedures were followed throughout the research to protect participants' rights and well-being.
2. Pilot Testing of Questions: The pre-prepared questions for the focus group sessions were pilot tested for clarity, relevance, and effectiveness in eliciting meaningful responses, involving a mock group to refine and ensure their suitability for the discussions.
3. Participant Information Leaflet: Prior to the focus group discussions, participants received a comprehensive information leaflet detailing the study's purpose, participation expectations, confidentiality measures, and withdrawal procedures.
4. Consent Form: Before participating in the focus group discussions, participants were required to review and sign an informed consent form, which outlined the study procedures, risks and benefits of participation, and their rights, including the option to withdraw at any time.

5. Focus Group Discussions: Focus group discussions were held across three university campuses, each targeting a specific discipline, where a moderator used a structured format and probing questions to guide the conversation and delve deeper into specific topics.

6. Data Transcription: Audio recordings of the focus group sessions were transcribed using Team software, ensuring efficient, accurate, and complete transcription of the discussions.

7. Data Cleaning and Verification: A data cleaning process was conducted to enhance the data's reliability, involving a thorough review and correction of the transcribed data against the original audio recordings to ensure accuracy and trustworthiness. In the conducted study, the researcher adhered to a five-step process for qualitative data analysis as outlined by Clarke & Braun, (2021):

   a. Familiarization with the Dataset: The researcher engaged in extensive reading and re-reading of the transcripts to become thoroughly familiar with the content. This process involved making detailed notes to form initial analytical observations. In line with Silverman, (2013)'s recommendations, audio recordings were also utilized. These recordings were crucial in understanding the tone and emotion of the speakers, group dynamics, and non-verbal cues such as pauses, sighs, changes in speech pace, and laughter. They were essential in ensuring the accuracy of the transcriptions and in revealing subtle aspects of speech, such as hesitations, emphasis on specific words, or changes in tone, which were key to grasping the nuances of the participants' responses.

   b. Coding: During this phase, the researcher generated succinct labels to capture significant ideas, concepts, explanations, examples, and metaphors relevant to the research questions. The use of audio recordings aided in refining codes that were not apparent in the transcript alone.

   c. Generating Initial Themes: The researcher examined the codes and collated data to develop broader patterns of meaning, which led to the formation of potential themes.

   d. Developing and Reviewing Themes: The researcher reviewed the established codes and themes to ensure that they presented a coherent narrative in line with the research questions.

   e. Refining, Defining, and Naming Themes: Lastly, the researcher evaluated the scope and focus of each theme. This step was crucial to ensure that the names of the themes were both informative and reflective of their content.

8. Thematic Analysis: Thematic analysis was used to systematically code the transcribed data, extract meaningful patterns and themes, and develop a comprehensive understanding of the concepts and issues emerging from the focus group discussions (Braun et al., 2019).

Each step of this procedure was designed to optimize the quality of the collected data and the insights derived from it. By adhering to a rigorous data collection procedure, this study ensured the quality, reliability, and validity of the data gathered through the focus group discussions. The combination of pilot testing, a participant information leaflet, a consent form, careful transcription, data cleaning, and thematic analysis provided a robust foundation for analyzing the complex relationship between social media usage and academic outcomes.

3.3 Participant Recruitment and Data Collection

At each campus, research assistants proactively invited students to participate in the focus group discussions. This direct approach ensured that the participants were well-informed about the purpose of the study and voluntarily consented to share their insights. To gather in-depth insights into students' perceptions and experiences regarding social media usage and its impact on academic outcomes, three focus group discussions were conducted. Each focus group consisted of 5 to 6 participants and lasted for approximately 45 to 54 min. The total number of participants across the three focus groups was 16, providing a rich and diverse range of perspectives for analysis. The focus group discussions were carefully guided by a moderator to ensure that all participants had the opportunity to share their thoughts and experiences. The discussions were audio-recorded and transcribed to facilitate further analysis and interpretation of the data.

Selecting participants for focus groups comes with inherent limitations. One significant limitation is the potential lack of representativeness owing to the small group size, making it challenging to generalize the findings. Additionally, the dynamics within the group and the potential for peer pressure can significantly influence individual responses, potentially skewing the data. The format of focus groups inherently limits the depth of each participant's engagement, offering fewer opportunities for a thorough exploration of individual perspectives or experiences. Another critical challenge is striking a balance between creating a diverse group to capture a broad range of perspectives and maintaining a homogeneous group to ensure focused and relevant discussions.

4 Findings

The findings presented the theme developed from the data. The main four identified themes are social media effect of attention and motivation, collaboration work and social media, distraction on academic work and reasons for the choices of social media.

4.1 Attention and Collaboration

Across the focus group discussions, a consistent theme emerged regarding the perceived impact of social media use on students' attention and motivation. Participants generally expressed the belief that their social media usage was not excessive or disruptive to their academic performance. They acknowledged the potential for distractions posed by social media, but they also emphasized their ability to manage their time effectively and balance social media engagement with their academic responsibilities. One research participant expressed the themes as follows: 'I use social media all the time, but it does not affect my academic performance.' Another research participant expressed the same line of argument 'I have been informed that my use of social media is excessive, and will impact on my attention on my academic work but my academic performance and result is not affected by it'.

Several participants noted that they utilized social media platforms for academic purposes, such as accessing study materials, connecting with classmates, and seeking clarification from instructors. They perceived these uses of social media as beneficial and complementary to their traditional learning methods.

While some participants acknowledged occasional instances where social media had temporarily diverted their attention from academic tasks, they did not attribute these distractions to any inherent negative effects of social media. Rather, they viewed them as lapses in self-regulation that could be addressed through improved time management strategies and mindful social media usage habits.

4.2 Social Media and Collaboration for Academic Work

Social media has emerged as a transformative tool for fostering collaboration and enhancing student engagement in group projects. By leveraging the interactive and connective nature of social media platforms, educators can cultivate a dynamic learning environment that empowers students to collaborate effectively, achieve shared goals, and ultimately improve their academic performance.

Social media platforms provide a virtual space for students to connect with peers, exchange ideas, and work together on projects seamlessly. This collaborative approach encourages students to actively participate in their learning, share their unique perspectives, and learn from their peers' diverse experiences. The ability to engage in real-time discussions, share files, and provide feedback seamlessly fosters a sense of shared responsibility and ownership of the project, motivating students to contribute their best efforts.

Social media platforms also facilitate asynchronous communication, allowing students to work on projects at their own pace and revisit discussions at their convenience. This flexibility is particularly beneficial for students with diverse schedules or time zone differences. Additionally, social media platforms enable students to ask questions and seek clarification from their peers or instructors outside of class time, promoting a continuous learning process.

Furthermore, social media platforms offer a range of features that streamline project management and enhance collaboration. The ability to share documents, schedule meetings, and engage in live video or audio discussions makes it easier for students to coordinate their efforts and clarify difficult concepts. This digital workspace fosters a sense of community and shared purpose, enabling students to work together effectively towards achieving their project goals.

4.3 Choice of Social Media Platforms

Students exhibit distinct preferences for social media platforms based on the purpose of use and the social circles they wish to engage with. Facebook, for instance, is often used to maintain connections with family and close friends. Some students may even restrict their Facebook usage to exclusively interact with family members, maintaining a separation between personal and academic circles. One research participant expressed this sentiment as follows: 'We are told to use different platforms for different modules. Imagine switching from one platform to another, receiving messages and alerts and reminder. In my group work we decided to use WhatsApp for all modules, simple and convenient."

Platforms like TikTok and Instagram, on the other hand, are primarily used for entertainment purposes. Students gravitate towards these platforms for their engaging content, such as short videos, memes, and music. This preference reflects the desire for relaxation and escape from academic pressures.

Interestingly, students often maintain separate social media identities for different social groups. Those who connect with colleagues and potential employers on LinkedIn may not include these connections on Facebook, WhatsApp, or Twitter. This distinction suggests a conscious effort to manage privacy and maintain different personas across social media platforms. Another student expressed the sentiment as follows: 'I have WhatsApp group with my colleagues, which is extensively used to ask questions, share ideas and documents. I don't want to use another platform, each lecture wants us to use what they prefer, I found it not efficient for my study.'

Further research could be conducted to delve deeper into the categorization of social media groups and explore the underlying motivations behind students' platform preferences. This research could shed light on the complexities of social media usage and provide valuable insights for educators and policymakers in tailoring interventions and strategies that align with students' social media behaviors.

5 Discussions

The discussion presented in this paper is organised around the key research questions that guided the study. It delves into the research findings, carefully examining them in the context of the existing literature. This approach ensures that the research findings are not merely presented in isolation but are instead situated within the broader landscape of knowledge on the topic.

5.1 How Does the Frequency and Purpose of Social Media Use Influence Students' Academic Attention, Motivation Influenced by Self-Regulation?

While research has established a link between excessive social media engagement and negative academic outcomes (Giunchiglia et al., 2018), determining the precise amount of social media usage that constitutes "too much" remains a challenge. Individual factors such as self-regulation skills, time management strategies, and social media usage habits play a significant role in mediating the impact of social media on academic performance.

Students with strong self-regulation skills and effective time management strategies may be able to manage their social media usage without experiencing significant distractions or declines in motivation. Conversely, students who struggle with self-regulation or have difficulty managing their time may find that even moderate levels of social media usage can negatively impact their academic performance.

Furthermore, the definition of "excessive" social media usage may vary depending on the individual student's academic workload, extracurricular activities, and personal commitments. For some students, even a few hours of social media usage per day may be too much, while others may be able to manage their time effectively and use social media responsibly without compromising their academic performance.

In light of these limitations, it is crucial for educators and policymakers to adopt a nuanced approach to addressing the impact of social media on student learning. Rather than focusing on arbitrary time limits or blanket restrictions, they should emphasize the development of self-regulation skills and mindful social media usage habits. By empowering students to make informed choices about their social media engagement, they can foster a learning environment that promotes academic success in the digital age.

While some students perceive their social media usage as neutral or even beneficial, a growing body of research indicates that excessive social media engagement can negatively impact academic performance, reduce attention, and diminish motivation to focus on academic work. The constant interruptions, the urge to stay connected, and the fear of missing out can all contribute to decreased focus, diminished motivation, and impaired productivity (Giunchiglia et al., 2018).

However, it is important to note that the impact of social media on academic outcomes is complex and varies depending on individual factors such as self-regulation skills, time management strategies, and social media usage habits. Students who can effectively manage their social media usage and prioritize academic tasks may experience less negative impact on their academic performance.

Therefore, educators and policymakers should focus on developing strategies to help students manage their social media usage, foster self-regulation skills, and promote mindful social media engagement. By empowering students to make informed choices about their social media usage, we can help them optimize their academic success in the digital age.

5.2 How Does Social Media Facilitate Collaboration Among Students, Including Group Project Coordination, Knowledge Sharing, and Peer-To-Peer Learning, and How Does This Collaboration Impact Students' Academic Outcomes?

Social media has emerged as a transformative tool for fostering collaboration and enhancing student engagement in group projects, offering a versatile platform that promotes knowledge sharing, facilitates communication, and streamlines project management. By leveraging the interactive and connective nature of social media platforms, educators can cultivate a dynamic learning environment that empowers students to collaborate effectively, achieve shared goals, and ultimately improve their academic performance.

Social media platforms provide a virtual space for students to connect with peers, exchange ideas, and work together on projects seamlessly. This collaborative approach encourages students to actively participate in their learning, share their unique perspectives, and learn from their peers' diverse experiences. The ability to engage in real-time discussions, share files, and provide feedback seamlessly fosters a sense of shared responsibility and ownership of the project, motivating students to contribute their best efforts.

While social media platforms offer a convenient and accessible space for collaboration among students, some students may prefer to exclude instructors from their group project discussions. This preference stems from a desire to maintain a sense of autonomy and privacy during the collaborative process. Students may fear that involving the instructor could lead to excessive scrutiny, judgment, or interference with their chosen approaches and decision-making.

Instructors' presence in the group's social media discussions can create an atmosphere of surveillance, diminishing students' sense of ownership and responsibility for the project. Students may feel hesitant to express their ideas freely or engage in open discussions if they perceive that their every word and action are being monitored and evaluated by the instructor. Moreover, students may worry that involving the instructor in their social media discussions could disrupt the group's dynamics and hinder their ability to work effectively together.

5.3   What Factors Influence Students' Preferences for Specific Social Media Platforms for Academic Purposes?

Social media platforms cater to diverse purposes, reflecting students' multifaceted digital lives. While platforms like Facebook and WhatsApp serve as primary channels for social connection and communication, platforms like TikTok and Instagram are primarily used for entertainment and relaxation. This distinction highlights the role of social media in fulfilling various needs, ranging from maintaining social ties to seeking escapism from academic pressures.

The preference for entertainment-focused platforms like TikTok and Instagram underscores the desire for leisure and distraction amidst academic demands. These platforms offer a respite from academic rigor, allowing students to unwind, engage with lighthearted content, and connect with others on a more informal level. The short-form videos, memes, and music on these platforms provide a quick dose of entertainment and humor, catering to students' need for relaxation and stress relief.

Furthermore, the conscious maintenance of separate social media personas for different social groups indicates an understanding of the varying expectations and norms across different online communities. Students recognize the importance of managing their digital presence and tailoring their online interactions to align with the specific context of each platform. This demonstrates a growing awareness of the implications of social media usage and the need to navigate the digital landscape responsibly.

The diverse usage patterns of social media among students reflect the multifaceted nature of these platforms and the varied needs they fulfill. Whether used for social connection, entertainment, or professional networking, social media has become an integral part of students' digital lives, offering a range of opportunities for communication, engagement, and self-expression.

6   Conclusion

While some students perceive their social media usage as neutral or even beneficial, a growing body of research indicates that excessive social media engagement can negatively impact academic performance, reduce attention, and diminish motivation to focus on academic work. The constant interruptions, the urge to stay connected, and the fear of missing out can all contribute to decreased focus, diminished motivation, and impaired productivity.

To effectively manage social media distractions and optimize academic outcomes, educators should adopt strategies that promote self-regulation skills, mindful social media usage habits, and effective time management strategies. Instead of imposing arbitrary time limits or blanket restrictions, educators should foster a collaborative learning environment that empowers students to make informed choices about their social media engagement.

Furthermore, when incorporating social media into collaborative work, educators should consider students' preferences for specific platforms and the varying expectations and norms across different online communities. Allowing students to choose the platforms that align with their preferences and the specific context of the collaborative task can enhance their comfort, engagement, and overall effectiveness in collaborative learning.

References

1. Al-Adwan, A.S., Albelbisi, N.A., Aladwan, S.H., Al Horani, O.M., Al-Madadha, A., Al Khasawneh, M.H.: Investigating the impact of social media use on student's perception of academic performance in higher education: evidence from Jordan. J. Inf. Technol. Educ. Res., 19 (2020). https://doi.org/10.28945/4661
2. Al-rahmi, W.M.: The impact of social media use on academic performance among university students: a pilot study. J. Inf. Syst. Res. Innov. (2013). http://seminar.utmspace.edu.my/jisri/
3. Al-Rahmi, W.M., Othman, M.S., Musa, M.A.: The improvement of students' academic performance by using social media through collaborative learning in malaysian higher education. Asian Social Sci. **10**(8) (2014). https://doi.org/10.5539/ass.v10n8p210
4. Al-Rahmi, W., Othman, M., Science, M.M.-A.S.: The improvement of students' academic performance by using social media through collaborative learning in Malaysian higher education. Academia.EduWM Al-Rahmi, MS Othman, MA MusaAsian Social Science, 2014 academia.Edu, **10**(8) (2014). https://doi.org/10.5539/ass.v10n8p210
5. Ansari, J.A.N., Khan, N.A.: Exploring the role of social media in collaborative learning the new domain of learning. Smart Learn. Environ. **7**(1) (2020). https://doi.org/10.1186/S40561-020-00118-7
6. Barton, B.A., Adams, K.S., Browne, B.L., Arrastia-Chisholm, M.C.: The effects of social media usage on attention, motivation, and academic performance. Act. Learn. High. Educ. **22**(1), 11–22 (2018). https://doi.org/10.1177/1469787418782817
7. Bhandarkar, A.M., Pandey, A.K., Nayak, R., Pujary, K., Kumar, A.: Impact of social media on the academic performance of undergraduate medical students. Med. J. Armed Forces India **77**(Suppl 1), S37 (2021). https://doi.org/10.1016/J.MJAFI.2020.10.021
8. Braun, V., Clarke, V., Hayfield, N., Terry, G.: Thematic analysis. Handbook of Research Methods in Health Social Sciences (2019). https://doi.org/10.1007/978-981-10-5251-4_103
9. Chukwuere, J.E.: Understanding the impacts of social media platforms on students' academic learning progress. Rev. Int. Geograph. Educ. **11**(9), 2671–2677 (2021). https://www.rigeo.org
10. Clarke, V., Braun, V.: Thematic analysis: a practical guide. London: SAGE (2021). https://uk.sagepub.com/en-gb/eur/thematic-analysis/book248481#description
11. Giunchiglia, F., Zeni, M., Gobbi, E., Bignotti, E., Bison, I.: Mobile social media usage and academic performance. Comput. Hum. Behav. **82**, 177–185 (2018)

12. Hassell, M.D., Sukalich, M.F.: A deeper look into the complex relationship between social media use and academic outcomes and attitudes. Inf. Res. Int. Electron. J. **21**(4) (2016)
13. Khan, N.A., Khan, A.N., Moin, M.F.: Self-regulation and social media addiction: a multi-wave data analysis in China. Technol. Soc. **64**, 101527 (2021). https://doi.org/10.1016/j.techsoc.2021.101527
14. Kolhar, M., Kazi, R.N.A., Alameen, A.: Effect of social media use on learning, social interactions, and sleep duration among university students. Saudi J. Biol. Sci. **28**(4), 2216 (2021). https://doi.org/10.1016/J.SJBS.2021.01.010
15. Silverman, D.: Doing Qualitative Research: A Practical Handbook, SAGE Publications. Doing Qualitative Research: A Practical Handbook. SAGE Publications, 488 (2013)
16. Thompson, P.: Communication technology use and study skills. Act. Learn. High. Educ. **18**(3), 257–270 (2017). https://doi.org/10.1177/1469787417715204

Information Technologies
in Radiocommunications

Radio Coverage Analysis for a Mobile Private Network

Marius-George Gheorghe, Vlad-Stefan Hociung, Alexandru Martian[✉],
and Marius-Constantin Vochin

National Science and Technology University Politehnica of Bucharest,
Bucharest 060042, Romania
{marius.gheorghe1303,alexandru.martian,marius.vochin}@upb.ro,
vladstefan.hociung@stud.etti.upb.ro
https://upb.ro/en

Abstract. In recent years, together with advances in the software defined radio (SDR) field, the concept of mobile private networks (MPNs) gained momentum. Several solutions based on open-source software modules and using SDR platforms as hardware were developed. In the current paper, the radio coverage of a private 4G mobile network was evaluated using the HTZ Communication software and field-test measurements. The network was implemented using an USRP B210 SDR platform and the srsRAN-4G software suite. A comparison between the coverage estimated using different propagation models and measurements is performed, together with an analysis of the obtained results.

Keywords: radio coverage · mobile private network · software defined radio · 4G · srsRAN · HTZ Communications

1 Introduction

Mobile communications networks, whose evolution started in the 1980 s with the first generation, can currently offer a wide range of services, transfer data rates in the order of tens of Gbps, latency as low as 1ms or less and can accommodate very large number of clients [1].

In recent years, the classical approach of using proprietary hardware and software for implementing different layers of the mobile communications networks was challenged, together with the development of open-source software suites and software defined radio (SDR) platforms [2,3]. Using such an approach, mobile private networks (MPNs) can be implemented, with application in different activity areas, like transportation, manufacturing, logistics, agriculture and education. The advantages offered by MPNs are related to the superior level of control, privacy and flexibility, when compared to commercial cellular networks.

Few papers in the literature discussed aspects related to radio coverage of mobile private networks. In [4], the authors use an SDR platform to deploy a Long Term Evolution (LTE) network testbed and evaluate an algorithm to predict the signal quality of a link between an User Equipment (UE) and an Evolved

Á. Rocha et al. (Eds.): WorldCIST 2024, LNNS 988, pp. 199–208, 2024.
https://doi.org/10.1007/978-3-031-60224-5_21

Node B (eNB). The extension of the cell coverage is analyzed numerically and experimentally in [5], in order to reach a desired cell size such as micro and macro-cells.

The main contributions of the current paper are related to the analysis of the radio coverage for a 4G mobile private network implemented using an USRP B210 SDR platform [6] as hardware and the srsRAN 4G suite [7] as software. The HTZ Communications [8] radio network planning and optimisation solution is used for estimating the radio coverage of the MPN using different propagation models. The obtained estimations are validated by comparison to field measurements, performed using a commercial mobile phone as UE.

The rest of the paper is structured as follows. In Sect. 2, the propagation models that will be used for estimating the radio coverage of the mobile private network are detailed. Section 3 contains a description of the setup that was used for implementing the mobile private network, including comments regarding its limitations. An analysis of the coverage estimated using the different propagation models and a comparison with measured data is included in Sect. 4. Section 5 concludes the paper and presents future research directions.

2 Propagation Models

In this section, several propagation models suitable for estimating the radio coverage of the mobile private network are detailed.

2.1 The ITU-R 1225 Propagation Model

ITU-R M.1225 Propagation Model [9] describes the guidelines for both the procedure and the criteria to be used in evaluating radio transmission technologies (RTTs) for a number of test environments. It ensures that the overall IMT-2000 objectives are met. There are 3 assumptions that are considered for this model [10]:

- for outdoor channels, if the mobile station (MS) antenna and the angular power spectrum are omnidirectional in the horizontal plane and multipath components (MPCs) arrive in the horizontal plane, then the Doppler power spectrum will have a "bathtub" shape;
- at the BS, the received MPCs arrive in a limited azimuth angular range while for indoor channels, a very large number of receive MPCs arrive uniformly distributed in elevation and azimuth for each delay interval at the BS;
- a short or half-wave vertical dipole is used as antenna.

2.2 The ITU-R 528-3 Propagation Model

ITU-R 528-3 propagation model [11] describes a prediction method suitable for aeronautical mobile and radio navigation services in the frequency range 125 - 15500 MHz. This model uses an interpolation method on basic transmission loss data from sets of curves (ground-air, ground-satellite, air-air, air-satellite, and satellite-satellite links). The parameters that have to be known for this model are: the operating frequency, the distance between the transmitter and receiver antennas, the heights of the antennas above mean sea level and the time percentage. Furthermore, if extra parameters are known (the transmitted power, the gain of transmitting/receiving antenna), an estimation for at least 95% of the time regarding the expected protection ratio or wanted-to-unwanted signal ratio exceeded at the receiver can be determined by this propagation model.

2.3   The ITU-R P.2001-4 Propagation Model

ITU-R P.2001-4 propagation model [12] was developed especially for wide-range communications (30 MHz–50 GHz) and has very high accuracy for a distance between 3–1000 km and antennas heights above ground level. The model provides a method for predicting basic transmission loss (fading and enhancements of signal level). If we consider the case of distances that are lower than 3 km, factors like clutter (buildings, vegetation, etc.) may have a high impact. This effect can be neglected if the antenna heights ensure an unobstructed path between transmitter and receiver. The model is based on the notion of combining seven sub-models representing seven separate propagation mechanisms (diffraction, ducting, troposcatter, sporadic-E, gaseous absorption, precipitation attenuation and multipath). From the above mentioned mechanisms, the first four provide end-to-end paths from transmitter to receiver. A different path through the atmosphere is followed by the radio waves for each of these principles, and all these contribute to the received signal. The last mechanisms don't provide new signal paths, but also affect the propagation of the signal between transmitter and receiver.

2.4   The ITU-R 1546-6 Propagation Model

ITU-R 1546-6 propagation model [13] is a method for predicting point-to-area radio propagation for terrestrial services in the frequency range of 30 MHz to 4000 MHz. This method depends on different factors such as: distance, antenna height, frequency and percentage time. The propagation curves represent field-strength values for 1kW effective radiated power and include land, sea, and mixed path at nominal frequencies of 100, 600 and 2000 MHz, respectively, as a function of various parameters; some curves refer to land paths, others refer to sea paths. The main characteristics of the propagation model in Rec. ITU-R P.1546-6 are: path type (terrestrial), distance (1 km–1000 km), antenna heights used for curves (10 m–3000 m), percentage of time 1%–50%, so on. For this propagation models is not mandatory to use a particular polarization.

2.5   The ITU-R 525/526-11 Propagation Model

The ITU-R 525/526-11 propagation model [14] is based on the phenomenon of diffraction. This is produced only by the surface of the ground or other obstacles, account must be taken of the mean atmospheric refraction on the transmission path to evaluate the geometrical parameters situated in the vertical plane of the path (angle of diffraction, radius of curvature, height of obstacle). Diffraction of radio waves over the Earth's surface is affected by terrain irregularities. Depending on the numerical value of the parameter of Obstacle surface smoothness criterion (see Recommendation ITU-R P.310) used to define the degree of terrain irregularities, three types of terrain can be classified: Smooth terrain, Isolated obstacles, Rolling terrain. Therefore, obstacles can be divided in: single knife-edge obstacles, single rounded obstacles, double isolated edges or multiple isolated cylinders, etc.

2.6   The M.2412 UMa Propagation Model

This model is the newest from the set of models used for estimating radio coverage in HTZ Communications [16]. ITU-R 56 defines a new term "IMT-2020" applicable to those systems, system components, and related aspects that provide far more enhanced capabilities. The capabilities of IMT-2020 include: very high peak data rate; very high and guaranteed user experienced data rate; quite low air interface latency; quite high mobility while providing satisfactory quality of service; enabling massive connection in very high density scenario; very high energy efficiency for network and device side; greatly enhanced spectral efficiency; significantly larger area traffic capacity; high spectrum and bandwidth flexibility; ultra high reliability and good resilience capability; enhanced security and privacy. The range of frequency for this propagation model is 0.5–100 GHz.

3   Implementation of the Private Mobile Network

The mobile private network whose radio coverage was evaluated in the current paper was implemented using the open-source srsRAN-4G framework [7] as software and an USRP B210 SDR platform [6] as hardware. Figure 1 presents a block diagram of the implemented MPN.

The software modules used for implementing the core part (srsepc) and the radio access part (srsenb) of the network ran on a host notebook, using Ubuntu 20.04 as operating system. The USRP B210 SDR platform, used as RF frontend, was connected to the host PC using an USB 3.0 connection, allowing a throughput large enough for accommodating a stable connection for an instantaneous bandwidth of 10 MHz (corresponding to a number of 50 Physical Resource Blocks (PRBs)). In order to be able to transmit with a larger transmit power for extending the coverage of the network, an external power amplifier (Mini-Circuits ZHL-2W-63-S+ [17]) was inserted between the transmit port of the USRP platform and the transmit antenna. A discone wideband antenna (Sirio SD-3000) was used as transmit antenna and a dipole antenna (Ettus Research VERT2450) was used as receive antenna.

A OnePlus Nord CE CG commercial-off-the-shelf (COTS) mobile phone was used as User Equipment (UE). Table 1 describes the parameters that were used to configure the radio access part of the implemented mobile private network.

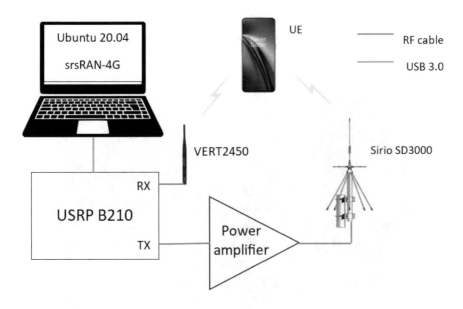

Fig. 1. Block diagram of the implemented mobile private network

The methodology used for obtaining the results presented in the following section consisted in the following steps:

- estimate the radio coverage of the implemented MPN in the HTZ Communications software, using 6 different propagation models;
- measure the received signal level using an UE and compare it with the 6 previously obtained coverage estimations.

The area that was chosen for performing the radio coverage analysis is located in the Nautical Based of the Constanta Maritime University, in the northern part of the city of Constanta, Romania. The area was chosen as it contains several containers somehow similar to the ones that are used in maritime transport,

Table 1. Parameters for the implemented mobile private network

Parameter	Value
LTE Band	7
Downlink Frequency	2680 MHz
Downlink EARFCN	3350
Tx Gain	80
Rx Gain	40
No. of PRBs	50
Bandwidth	10 MHz

which is the focus of the Solid-B5G project [18], which funded this research. Figure 2 shows the surroundings of the area where the mobile private network was installed and the measurements were performed.

Fig. 2. Surroundings of the area where the measurements were performed (Nautical Base of the Constanta Maritime University)

3.1   Limitations of the Implemented MPN

Several aspects related to the performed implementation should be mentioned as limitations. As the SDR platform which was used as RF front-end is not a calibrated equipment, the RF output power considered for performing the coverage estimations (−5 dBm) was measured using a spectrum analyzer and might be not very accurate. Moreover, even using a power amplifier, the power level obtained is far below that offered by commercial mobile networks. As a consequence, the area that can be covered by the implemented MPN is not very large and, because of that, the measurement route that could be followed is not a very long one.

4   Results and Analysis

In this section, the results that were obtained will be presented, including both estimations that were achieved using the HTZ Communications software [8] and measurements that were performed using a COTS mobile phone as UE.

The six different propagation models that were detailed in Sect. 2 were used for estimating the radio coverage of the mobile private network, the transmitter site was configured using the parameters mentioned in Table 1, and for the receiver site an antenna height of 1.2 m was considered. The threshold for displaying the received signal level was set at a value of −127 dBm.

Figure 3 presents the coverage maps that were obtained, calculated for a radius of 300 m around the location of the transmitter. It can be noticed that the most optimistic estimation was obtained for the ITU-R 1546-6 model, which indicated a Reference Signal Received Power (RSRP) above the threshold for the whole area included in the capture.

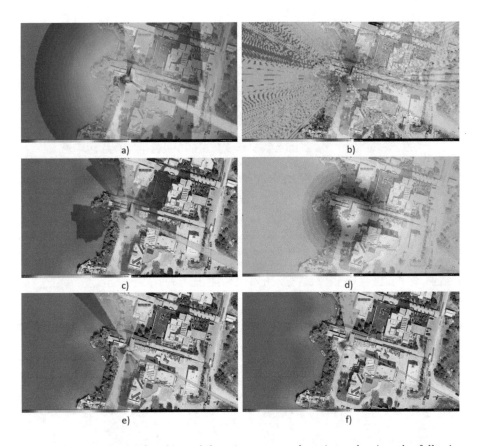

Fig. 3. Coverage maps for the mobile private network estimated using the following propagation models: a) ITU-R 1225 b) ITU-R 528-3 c) ITU-R 2001-4 d) ITU-R 1546-6 e) ITU-R 525/526-11 f) M.2412 UMa

In order to validate the obtained estimations, we performed outdoor measurements around the transmitter site, as it can be seen in Fig. 4.

The results of the correlation between the theoretical estimations and the experimental data, for the route described in Fig. 4, are given in Fig. 5. The green line represents the theoretical estimated RSRP, corresponding to one of the eight propagation models, and the yellow line represents the measured RSRP, using the followed route represented as the x axis. It can be noticed that the two

Fig. 4. Path that was used for performing the measurements, visualized in Google Earth Pro

models that produced the closest estimates, as compared to the measured data, are the ITU-R 1225 and ITU-R 528-3 models.

Table 2 includes average performance metrics for each of the six propagation models, including the percent for the followed measurement route for which the

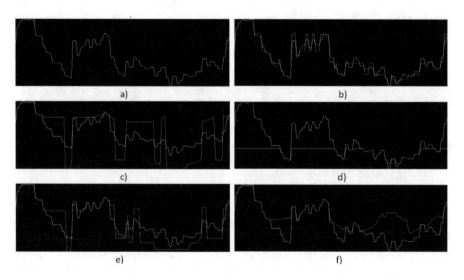

Fig. 5. Correlation between the estimation (green line) and the measurements (yellow line) for the following propagation models: a) ITU-R 1225 b) ITU-R 528-3 c) ITU-R 2001-4 d) ITU-R 1546-6 e) ITU-R 525/526-11 f) M.2412 UMa

difference between the estimated and measured RSRP was smaller than 6 dB, the standard deviation (in dB), the average error (in dB) and the correlation factor.

Table 2. Average performance metrics for the six propagation models

Propagation Model	<6 dB (%)	Std. dev. (dB)	Avg. err. (dB)	Correl. factor
ITU-R 1225	100	0.19	−0.04	1.00
ITU-R 528-3	100	2.45	−0.28	0.98
ITU-R 2001-4	32.85	12.43	0.31	0.8
ITU-R 1546-6	62.77	10.01	−7.67	0.71
ITU-R 525/526-11	53.28	7.10	−1.85	0.90
M.2412 UMa	69.34	6.95	5.35	0.83

It can be noticed that the best results are obtained for the ITU-R 1225 propagation model, for which the estimation overlaps almost perfectly over the measured data (average error of only −0.04dB, correlation factor 1). A very good performance is also obtained for the ITU-R 528-3 propagation model, with an average error of −0.28 dB and a correlation factor of 0.98. On the opposite end, the largest average error and the lowest correlation factor are obtained for the ITU-R 1564-6 propagation model.

5   Conclusion and Future Work

The radio coverage for a private 4G mobile network, implemented using the srsRAN 4G software suite and an USRP B210 SDR platform, was analyzed in the current paper. The HTZ Communications software was for obtaining different estimations for the network coverage, using different propagation models. The estimations were validated by means of comparison with measurements that were performed using a COTS mobile phone as UE. A very good correlation was obtained for two of the propagation models, ITU-R 1225 and ITU-R 528-3. As future work, we intend to perform further analyses in different frequency ranges and extend the experiments for private 5G mobile networks.

Acknowledgement. This research has been funded from NO Grants 2014-2021 under Project contract no. 42/2021, RO-NO-2019-0499 - "A Massive MIMO Enabled IoT Platform with Networking Slicing for Beyond 5G IoV/V2X and Maritime Services" - SOLID-B5G. The authors would also like to thank Prof. Razvan Tamas (Constanta Maritime University) for allowing us to perform the measurements in the Nautical Base of the Constanta Maritime University.

References

1. Navarro-Ortiz, J., Romero-Diaz, P., Sendra, S., Ameigeiras, P., Ramos-Munoz, J.J., Lopez-Soler, J.M.: A survey on 5G usage scenarios and traffic models. IEEE Commun. Surv. Tutorials **22**(2), 905–929 (2020)
2. Eswaran, S., Honnavalli, P.: Private 5G networks: a survey on enabling technologies, deployment models, use cases and research directions. Telecommun. Syst. **82**(1), 3–26 (2023)
3. Arnaz, A., Lipman, J., Abolhasan, M., Hiltunen, M.: Toward integrating intelligence and programmability in open radio access networks: a comprehensive survey. IEEE Access **10**, 67747–67770 (2022). https://doi.org/10.1109/ACCESS.2022.3183989
4. Feitosa, W.D.O., da Silva, R.A., Monteiro, V.F., Cavalcanti, F.R.P.: RSRP prediction on LTE network testbed using a software defined radio (SDR) platform
5. Molla, D.M., Badis, H., George, L., Berbineau, M.: Coverage extension of software defined radio platforms for 3GPP 4G/5G radio access networks. In: 2021 13th IFIP Wireless and Mobile Networking Conference (WMNC) (pp. 55–62). IEEE (2021)
6. Ettus Research USRP B210 SDR platform (2023). https://www.ettus.com/all-products/ub210-kit/
7. srsRAN 4G Software Suite (2023). https://www.srsran.com/4g
8. HTZ Communications (2023). https://atdi.com/products-and-solutions/htz-communications/
9. Guidelines for evaluation of radio transmission technologies for IMT-2000 (2023). https://www.itu.int/dms_pubrec/itu-r/rec/m/R-REC-M.1225-0-199702-I!!PDF-E.pdf
10. Tataria, H., Haneda, K., Molisch, A.F., et al.: Standardization of propagation models for terrestrial cellular systems: a historical perspective. Int. J. Wireless Inf. Netw. **28**, 20–44 (2021)
11. Recommendation ITU-R P.528-3 (02/2012) (2023). https://www.itu.int/dms_pubrec/itu-r/rec/p/R-REC-P.528-3-201202-S!!PDF-E.pdf
12. Recommendation ITU-R P.2001-4 (09/2021) (2023). https://www.itu.int/dms_pubrec/itu-r/rec/p/R-REC-P.2001-4-202109-S!!PDF-E.pdf
13. Recommendation ITU-R P.1546-6 (08/2019) (2023). https://www.itu.int/dms..pubrec/itu-r/rec/p/R-REC-P.1546-6-201908-I!!PDF-E.pdf
14. Recommendation ITU-R P.526-11 (10/2009) (2023). https://www.itu.int/dms..pubrec/itu-r/rec/p/R-REC-P.526-11-200910-S!!PDF-E.pdf
15. Propagation model for Recommendation ITU-R P.2345-1 (2023). https://www.itu.int/dms_pub/itu-r/opb/rep/R-REP-P.2345-1-2016-PDF-E.pdf
16. Report ITU-R M.2412-0 (10/2017), Guidelines for evaluation of radio interface technologies for IMT-2020 (2023). https://www.itu.int/dms...pub/itu-r/opb/rep/R-REP-M.2412-2017-PDF-E.pdf
17. Mini-Circuits ZHL-2W-63-S+ power amplifier (2023). https://www.minicircuits.com/pdfs/ZHL-2W-63-S+.pdf
18. Solid-B5G Research Project (2023). https://solid-b5g.upb.ro/

Tunable Reflector/Absorber Surfaces for Next Generation Wireless Communication Systems

Sandra Costanzo and Francesca Venneri[✉]

DIMES – University of Calabria, 87036 Rende, CS, Italy
`venneri@dimes.unical.it`

Abstract. One of the main challenges in research beyond 5G and 6G networks is the fruition of smart propagation environments. Reconfigurable Intelligent Surfaces (RIS) represent a promising technology for realizing controllable wireless propagation environments that can actively cooperate in data transfer and processing. A reconfigurable PIN-loaded metamaterial unit-cell is investigated for next-generation communications, Device-to-Device systems and Internet of Things. The proposed cell allows to implement switchable reflector/absorber surfaces for different polarizations, to selectively allow/inhibit signal transmission with respect to the polarization of the impinging wave. Thanks to its features, the proposed metamaterial configuration could be very attractive for designing and customizing RIS prototypes for specific communication scenarios.

Keywords: RIS · Metamaterial absorbers · 5G and 6G

1 Introduction

Currently, research beyond 5G, or even sixth generation (6G), is underway to offer increasingly intelligent and ubiquitous connectivity [1]. Future wireless networks are expected to provide a smart, distributed platform for communications, sensing, and computing. To achieve this goal, the wireless environment will need to play an active role in the implementation of next generation communication systems. The propagation environment will be 'smart'; it will become a smart reconfigurable space that actively cooperates in data transfer and processing.

In current wireless systems, the environment is instead out of control, it is a passive observer in the data exchange process. Furthermore, the environment usually has negative effects, such as uncontrollable interferences due to reflections and refractions, path-loss and fading phenomena.

Recently, a new technology called Reconfigurable Intelligent Surfaces (RIS) is proposed as a potential solution for realizing controllable wireless propagation environments. RISs, or smart skin/surfaces, are artificial surfaces having the abilities to manipulate the electromagnetic signal impinging upon them [2, 3]. RIS concept is based on the well-established concept of meta-surfaces [4], which are planar structures, made of smaller and periodically arranged conductive elements printed on a dielectric substrate. Meta-surfaces can be integrated with active elements and/or materials, such as

Á. Rocha et al. (Eds.): WorldCIST 2024, LNNS 988, pp. 209–215, 2024.
https://doi.org/10.1007/978-3-031-60224-5_22

varactor or PIN-diodes, to offer a variety of reconfiguration capabilities such as absorption, anomalous reflection, refraction, collimation, wave focusing towards a specified direction (beamforming), signal polarization, etc. (Fig. 1).

Due to the above highly attractive skills, RISs constitute a promising physical-layer technology for beyond 5G and 6G of mobile communications scenarios [5, 6], both in indoor as well as in outdoor settings (e.g. a RIS can cover the walls of a room or the facade of a building).

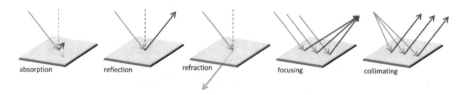

Fig. 1. Elementary functions supported by meta-surfaces.

In this paper, a preliminary investigation on a controllable RIS unit-cell is presented. The proposed cell allows to realize a switchable reflector/absorber meta-surface for different polarizations [7], which can be fruitfully exploited to selectively allow/inhibit signal transmission with respect to the polarization of the impinging wave.

The configuration could be adopted, for example, to provide additional transmission paths, when the line-of-sight (LOS) communication link is blocked, or to absorb/inhibit undesired signal reflections/refractions from large objects located along the path coming from the source. The proposed reflector/absorber unit-cell is composed by two sets of metallic resonant elements, printed on a grounded dielectric slab and loaded by a pair of PIN-diodes. Each set is able to independently manipulate the orthogonal polarizations of EM impinging waves. A preliminary and comprehensive numerical characterization of the proposed configuration is performed. The results achieved at this stage are very promising and provide the basis for designing and customizing a RIS prototype for a specific communication scenario.

2   Switchable Reflector/Absorber Meta-surface Unit Cell

2.1   Unit Cell Layout and Principle of Operation

The layout depicted in Fig. 2 is adopted to design a switchable reflector/absorber meta-material cell. The proposed unit cell is based on our previous design of a dual polarized reflectarray cell [8, 9]. It is composed by two alternately arranged pairs of linearly polarized H-shaped patches, operating at the same resonant frequency. They are rotated each other by 90°, in order to offer the ability to independently manipulate the orthogonal polarizations of EM impinging waves. Each patch is characterized by a beginning square element of dimensions L × L. A smaller SL × SL-square is removed from the center of the resonant sides of the beginning patch, where S is the scaling factor. Both L as well as SL, are properly chosen in order to fix the resonance at the desired operating frequency. As a matter of fact, the resonance frequency of each patch is inversely proportional to the effective resonant side length, namely $L^{eff} = (1 + S)L$ [10].

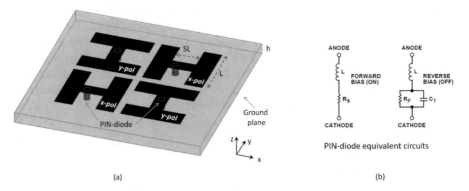

(a)   (b)

Fig. 2. Metamaterial unit cell: (a) layout of a PIN-loaded reflector/absorber unit cell; (b) PIN diode equivalent circuits.

At resonance, each pair of patches strongly couples to the electric field component of the incident wave parallel to the central metallic strip (see Fig. 2(a)), thus providing an independent electric response.

In the unit cell version proposed in this work (Fig. 2(a)), the elements are loaded with a shunt-connected PIN-diode, whose position is suitably chosen to allow the desired reflector/absorber behavior, by varying the applied bias voltage. Indeed, PIN diodes are usually adopted as current-controlled resistors at RF and microwave frequencies, with resistances that can range from a fraction of an ohm, when forward biased, or ON, to greater than 10 kΩ, when reverse biased, or OFF [11].

In particular, when a reverse- or zero-bias voltage is applied (Fig. 2(b)), the PIN-diode appears as a large resistance, R_P, shunted by a capacitance, C_T, thereby its equivalent circuit appears as an open circuit. In this case, the patch behaves as a reflectarray element, reflecting back the impinging wave with very low reflection losses. On the other hand, by tuning the bias voltage from zero bias to forward bias the PIN-diode behaves like a resistance (R_S in Fig. 2(b)), namely the effective ON resistance of the diode. In this case, a current flows through the diode. By properly adjusting the bias voltage and the diode position, the above state can be exploited to achieve the perfect absorption condition [12], namely to obtain the matching between unit cell and free space impedances at the design frequency (namely $Z_{\text{cell}}(f) \cong \zeta_0 = 376\Omega$, i.e. $\Gamma(f) \cong 0$).

2.2 Unit Cell Characterization

In order to give a preliminary validation of the proposed structure, a 6 GHz cell is designed and simulated. A Diclad880 substrate ($\varepsilon r = 2.24$, thickness h = 1.524mm) is considered and a periodicity equal to $\Delta x = \Delta y = 0.4\lambda$ at 6 GHz is fixed. The analysis is performed by adopting the infinite array approach (Ansys Designer) and assuming a normally incident plane wave [13]. Both x-polarization (TM) and y-polarization (TE) are considered. The cell is designed by following the design rules outlined in [8–10]. The sizes of the H-shaped patches are equal to (L - S) = (9 mm - 0.39) while the PIN-diode is placed at a distance of 1 mm from the center of each patch. The PIN diode model used for the simulations is Skyworks APD Series (ON resistance 2.5Ω, capacitance 0.3 pF). The switching capability of the cell between the reflection/absorption behaviors are investigated in Fig. 3, which shows the reflection coefficient of the cell computed for both TM and TE polarizations when all the PIN-diodes are in OFF-state (Fig. 3(a)) or in ON-state (Fig. 3(b)). It can be observed the ability of the cell to switch from reflection function ($\Gamma(f) \cong 1$ in Fig. 3(a), i.e. $A(f) = 1 - |\Gamma(f)|^2$ very low) to absorption operation mode ($\Gamma(f) \cong 0$ in Fig. 3(b), i.e. $A(f) = 1 - |\Gamma(f)|^2$ high $\rightarrow 1$).

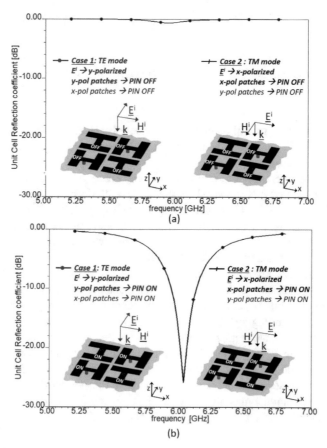

Fig. 3. Simulated reflection coefficient under normal incidence (both TE and TM modes) for different PIN-diodes configurations: (a) PIN-diodes OFF, (b) PIN-diodes ON.

Afterwards, the selective response of the proposed metamaterial cell to different polarized EM waves is investigated. As discussed in previous section, the two pairs of orthogonally oriented patches (Fig. 2(a)) separately react to the orthogonally polarized incident waves. This feature allows us to control the reflection or absorption of certain polarized incident wave, by properly biasing the PIN-diodes in the corresponding patches.

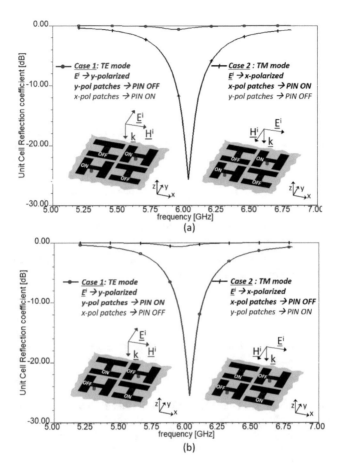

Fig. 4. Simulated reflection coefficient under normal incidence (both TE and TM modes) for different PIN-diodes configurations: (a) x-pol patches → PIN ON, y-pol patches → PIN OFF; (b) x-pol patches → PIN OFF, y-pol patches → PIN ON.

As shown in Fig. 4(a), if we consider the TM-mode, the reflection coefficient undergoes a minimum around 6 GHz, when the diodes mounted in the x-polarized patches are forward biased (ON) and those in y-polarized patches are reverse-biased (OFF). In other words, this means a nearly perfect absorption of the x-polarized wave. In the same Fig. 4(a), it can be observed that the reflection coefficient changes to almost unity (high reflection of the y-polarized wave) when the TE-mode is considered, since the y-polarized patches are in the OFF-state. Similar and opposite considerations can be made on the results depicted in Fig. 4(b).

In conclusion, the above results make the proposed structure very appealing for the design of RISs able to separately control the behavior of the orthogonally components of impinging waves. As future developments, the proposed structure will be further optimized and investigated in order to check and verify other crucial performances, such as angular and polarization stability of the unit cell, scalability, etc.

Furthermore, other general issues related to the integration of PIN diodes and biasing circuitry will be considered and addressed, both theoretically as experimentally.

3   Conclusion

Wireless environment challenges in research beyond 5G and 6G networks have been discussed. RISs, or smart skin/surface, concept has been addressed as the most promising solution to realize controllable and cooperative wireless propagation environments.

A PIN-loaded meta-surface unit-cell has been investigated for next-generations communications. A switchable reflector/absorber dual-polarized cell has been designed and simulated, to selectively allow/inhibit signal transmission with respect to the polarization of the impinging wave.

Thanks to its features, the proposed metamaterial configuration has revealed itself very attractive for designing and customizing RIS prototypes for specific communication scenarios. As future development, the configuration will be optimized also taking into account the issues related to the biasing network.

References

1. 5G Italy. The Global meeting in Rome (2021). https://www.5gitaly.eu/it/homepage/
2. Basar, E., Di Renzo, M., De Rosny, J., Debbah, M., Alouini, M., Zhang, R.: Wireless communications through reconfigurable intelligent surfaces. IEEE Access **7**, 116753–116773 (2019). https://doi.org/10.1109/ACCESS.2019.2935192
3. Renzo, M.D., Debbah, M., Phan-Huy, D.T., et al.: Smart radio environments empowered by reconfigurable AI meta-surfaces: an idea whose time has come. J. Wireless Commun. Network **2019**, 129 (2019). https://doi.org/10.1186/s13638-019-1438-9
4. Munk, B.A.: Frequency Selective Surfaces: Theory and Design. New York, NY, USA, Wiley Interscience (2000)
5. Di Renzo, M., et al.: Smart radio environments empowered by reconfigurable intelligent surfaces: how it works, state of research, and road ahead (2020). arXiv:2004.09352v1
6. Costanzo, S., Venneri, F.: Smart EM surfaces for future wireless communication systems. In: 34th General Assembly and Scientific Symposium of the International Union of Radio Science, URSI GASS (2021). https://doi.org/10.23919/URSIGASS51995.2021.9560262
7. Zhu, B., Fenga, Y., Zhao, J., Huang, C., Jiang, T.: Switchable metamaterial reflector/absorber for different polarized electromagnetic waves. Appl. Phys. Lett. **97**, 051906 (2010). https://doi.org/10.1063/1.3477960
8. Costanzo, S., Venneri, F., Di Massa, G.: Dual-polarized reflectarray cell for 5G applications. In: 33rd General Assembly and Scientific Symposium of the International Union of Radio Science, URSI GASS 2020, 9232345 (2020). https://doi.org/10.23919/URSIGASS49373.2020.9232345
9. Costanzo, S., Venneri, F., Borgia, A., Massa, G.D.: Dual-band dual-linear polarization reflectarray for mmWaves/5G applications. IEEE Access **8**, 78183–78192 (2020). https://doi.org/10.1109/ACCESS.2020.2989581
10. Costanzo, S., Venneri, F.: Miniaturized fractal reflectarray element using fixed-size patch. IEEE Antennas Wireless Propag. Lett. **13**, 1437–1440 (2014). https://doi.org/10.1109/LAWP.2014.2341032

11. Analog Devices Homepage. https://www.analog.com/en/analog-dialogue/articles/driving-pin-diodes-with-op-amps.html. Accessed 04 Dec 2022
12. Venneri, F., Costanzo, S., Di Massa, G.: Fractal-shaped metamaterial absorbers for multireflections mitigation in the UHF band. IEEE Antennas and Wireless Propagation Lett. **17**(2), 255–258 (2018). https://doi.org/10.1109/LAWP.2017.2783943
13. Venneri, F., Costanzo, S., Di Massa, G.: Bandwidth behavior of closely spaced aperture-coupled reflectarrays. Int. J. Antennas Propag. **11**, 846017 (2012). https://doi.org/10.1155/2012/846017

Author Index

Printed in the United States
by Baker & Taylor Publisher Services